D0896569

Editing Contributor: Chris Pepper
Cover Design: Onur Burc
Publisher: Tavo Reno Publishing
www.thedeliveryman.info

GRATEFUL GEEK

50 Years of Apple and Other Tech Adventures

By Jean-Louis Gassée

TABLE OF CONTENTS

INTRODUCTION

This book is…

First, a look back, in amazement and gratitude, to a very long procession of people who helped me, liked and even loved me sometimes, who smiled indulgently and tolerated me, taught me, and sometimes motivated me to do more; and more importantly, to be a better person. In the autumn of my years, I bow to them all, even to the ones I, unfortunately, disliked or failed to appreciate. I won't be able to acknowledge them all, but they live in me as I write this book.

Second, it is a paean to the personal computers that lit up my business life. In this book, you'll find my explanation for our fascination with the machines that give wings to our minds and bodies. They no longer dwell solely on our desks or in our bags; they inhabit our pockets and now our wrists; they help us think, communicate, organize, learn and play. I was lucky to enter real professional life at the start of the PC era. Over fifty years later, I remain amazed, occasionally frustrated, and always hopeful to see personal computers continue to delight us.

Third, 50 years in the world of tech provided many, many opportunities to make mistakes. I recount most, not in a confessional spirit, but more like a picaresque tale of one fumble or illusion after another, with a grateful smile for those who, like a small Inuit tribe, jumped with me from one ice floe to the next. I survived, after all, which feeds my sense of wonderment and gratitude for those who shared those adventures— or suffered from them. But no advice expressed or implied. We know the old joke about good judgment being the fruit of experience, and experience resulting from bad judgment.

I hope you'll enjoy my stories as much as I enjoyed living them— and that you have or will find your own. Either in you or around you.

VENI:
THE FORMATIVE
YEARS

The Making of a Geek

Turning Point 1:
Two Geeks in Brittany

A kind mentor helps me recover my balance and encourages my nascent passion for electronics.

At age 10, I lost my balance. Coming from the strictly disciplined environment of the Jules Ferry primary school in Joinville-le-Pont, a suburb east of Paris, I entered a nearby *lycée (high school)* that offered much looser supervision. This led to much truancy, bad grades – and entering the world of science-fiction books — a habit which would last over two decades. It didn't take long for my parents to figure things out. Unfortunately, perhaps because of their respective upbringings, they didn't know better than meting out verbal and occasionally physical punishment. I had become a "bad kid" and was expelled from the *lycée*. A professor of medicine friend of the family briefly hospitalized me for a detailed examination and pronounced me agitated and agile of mind, nothing seriously wrong.

A few months later, Bretonne family connections got me in front of a priest called Joseph Guellec, superior of the Kreisker Roman Catholic boarding school in Saint-Pol-de-Léon, a small town in the western region of France called Finistère (meaning Lands End). I easily passed an entrance exam, and that's when my luck turned: I met the Prefect of Discipline, a kind priest named Yves Kerdilès. Kind and geeky. When I told him I had already built a galena[1] radio,

---

[1] Galena is crystallized lead sulfide that works like an elementary semiconductor, hence its use in hobby radio sets.

he made it a habit of inviting me to his office after study hours so we could leaf through the 'radio' magazines of the time[2], peruse wiring diagrams, read ads that were as or more interesting than the articles and salivate looking at his latest acquisition: an OC 71 germanium transistor, not yet silicon, but the beginning of the semiconductor era.

Away from my troubled family, I prospered and immediately made the honor roll, much to my parents' relief— they were to divorce, to my own mix of sorrow and deliverance, a couple of years later. My mother fled to Spain, rejoined by my younger brother, and I stayed with my father.

Still, thanks to my parents, I got a solid combination of humanities and science education at the hands of dedicated and safe priests. Life was spartan: no central heating, no hot water in the communal bathroom, and a weekly visit to a collective shower room for a thirty-second stay under hot water. Humid Brittany winters made it difficult to keep our clothes dry. I still remember the joy of hot showers and dry underwear when I moved to the Lycée Hoche in Versailles after entering the college track.

At the Kreisker I followed the French Eagle Scout path, with its focus on practical training, physical endurance, and mental leadership. On the geeky side, I built clandestine radios with antennas hidden in baseboards and managed a secret darkroom to develop negatives from a Kodak Vest Pocket[3] camera a grandfather gave me. That camera didn't last long: I took it and other objects apart, including my own watch. Most of the time, being a born and impatient klutz (a lifelong trait), I failed to restore them to their original state — a recurring source of shame. Fortunately, the

---

[2] For older French readers: Le Haut-Parleur or Radio Plans.

[3] https://camerapedia.fandom.com/wiki/Vest_Pocket_Kodak

embarrassment never stopped me from pursuing the next project, a lasting drive that was to take me to many interesting places and people.

I also dabbled in explosives as it was easy at the time. It is now unbelievable how easy it was to buy sulfur; potassium perchlorate and hydrochloric, nitric, and sulfuric acids at a nearby store. This allowed me to entertain classmates during the boring mandatory Sunday afternoon walks to the nearby beach. Our physics teacher liked *my burning interest in chemistry*, and eventually gave me a key to the school's lab to conduct further experiments. I stopped when I realized I was on my way to building an unstable explosive: nitrocellulose.

The religious foundation of the school wasn't its best side. The good priests had the best intentions and motives for taking good care of us. And they were safe. We've all heard about the abuse perpetrated by clerics on young children. None of that happened to us there as far as we know. I still recall how a young and blushing priest who looked like he had just begun shaving came in and appeared to be attracted to some of us, a fact promptly circulating inside our little and always whispering community. Two months later, the unfortunate cleric was gone. Management paid attention and kept us safe — a memory that always comes back when I read stories of abuse by teachers, members of religious orders or not. Why didn't their management pay attention?

Not everything was so benign. There was confession, something I see now as an instrument of mental and physical coercion, creating and maintaining shame for the natural instincts of poor kids who had trouble with their emerging sexuality. No public, empathetic discussion, only references to hell, whispered talk in the confessional, and orders of penance. I felt relief when I left the Kreisker for a secular college — and painful thoughts came back when, as a requirement for my first marriage in a Roman Catholic parish, I had to present a mandatory "bill of confession" to the celebrant.

In the end, I passed the state high school exam (called *baccalauréat*) with honors and was naturally directed to what is known in France as the Grandes Écoles track, a path to prestigious engineering and science schools. I did moderately well there, including a side trip to a summer camp at the state civil aviation school, where I aced the written and oral exams, but flunked the final flight test. My perhaps lame excuse is that my father believed the school, which claimed previous flying experience wasn't required. In reality, every student but me came to the camp after extensive flight lessons. I was inconsolable then; the dream of becoming an airline pilot was gone! In retrospect, I was twice lucky. First, I probably didn't possess the innate skills and judgment to fly planes; I later proved this by destroying several cars and bending many fenders in my younger driving years. And second, from the plane's cabin, I now see more clearly what being an airline pilot has become and compare it negatively against my life's adventures.

As a consolation prize of sorts, I just barely got a junior math and physics degree from Orsay University near Paris.

I was nineteen years old when tragedy struck: my first love was killed in an accident at the wheel of her parents' car just days before we were to head out on a driving tour of Europe (with our parents' consent — and her sister as chaperone). At the moment I felt numb and didn't know how to process the event, a state that lasts to this day.

Turning Point 2:
The Apple Impact

I join Apple — almost by accident — and soon realize it is a bigger and better opportunity.

"I can get you shares in Apple's upcoming IPO."

"No thanks, I'm a repented mathematician, I don't play the stock market[4], I'll stay with my Tandem[5] shares."

"Hmmm…, as you wish."

The person offering me Apple IPO shares is Aaron Orlhansky, a noted analyst from the big Wall Street Research firm Gideon Gartner, now called The Gartner Group.

Aaron continues:

"I'm friends with Tom Lawrence, Apple's Europe VP, and will have lunch with him next week. I know he's looking for an executive to start Apple's French subsidiary. Would you know anyone he could hire?"

"Yes, me."

"What? But you have this great job running Exxon Information Systems' affiliate here."

---

[4] To underline my financial savvy, a December 1980 IPO share was priced at $22. Today, after 5 splits and innumerable feats, that share would be worth $31,308, a 144,581% increase (aside from inflation).

[5] Tandem, founded by people I knew from my old employer HP, once led the fault-tolerant computing market — since made obsolete by innovations in standard hardware and software.
https://en.wikipedia.org/wiki/Tandem_Computers

"No, no, no, I fell for Exxon's diversification into Information, the Oil of the 21st Century — and so did you and every other Wall Street seer. This whole tale of diversifying away from petroleum is a concoction of the prestigious Boston Consulting Group. I've seen the presentation. The idea is to progressively insulate the Standard Oil descendant from Middle East crises and to jump on the still young computer industry growth curve. It sounds great, visionary, sure to succeed with the giant's financial and managerial resources. But, no, trust me, Exxon is out of its mind and natural habitat. Their culture is wrong for the world of chips, software and the strange creatures who inhabit it. I've been there more than a year now and *know* it's hopeless. They're like members of an abstemious order gingerly trying to run a Paris nightclub."

From inside the Exxon Information Systems, I was sure their superficially sound diversification strategy would someday become another case study in Culture Eats Strategy for Breakfast.

Back in my HP days I played amateur kremlinologist for Aaron. He bought me lunch, and I decoded the willfully misleading PR material from companies Aaron dissected especially IBM, then at the height of its market power.

(For readers too young to remember, the grand old Soviet Union did its best to keep the workings of its high administration, the *nomenklatura*, to itself. Kremlinologists specialized in looking at October Revolution celebration pictures and figuring out why someone disappeared, moved up or down on the tribune, or suddenly emerged; they tried to make sense of the mixture of information and *dezinformatsiya* intended to confound the Enemies of the People.)

You see the parallel.

Besides dressing up unilateral decisions as "our customers demanded this", my favorite IBM hot air tactic was dismissing a competitive threat using the following three-step dismissal: It's Nothing, followed by You Don't Need It, ending with We'll Have It In Six Months.

My glee and experience disassembling corpospeak would again come to play a key role later in my Apple days. Successfully taking apart Steve Jobs' chancy Mac Office storytelling would trigger my move to the US.

A week later I met Tom Lawrence. We hit it off immediately. Nicknamed Lawrence of Europe for his charm and expansive lifestyle, Tom liked my combination of computer industry (HP, Data General) and management (two CEO stints) experience. I got the job. Only later did I see the benefit of coming late into a recruitment obstacle course. Another candidate had previously gone through the full process, providing useful calibration. When I came in, I was more easily and rapidly evaluated. As we'll see in a later chapter, this was an eerie, lucky repeat of the coming late in the recruitment process, which earlier got me my first and fundamental tech job at HP France. (Later I got to meet the gentleman Apple almost hired. He became the successful subsidiary CEO of another personal computer company, Compaq, and took his rejection with good humor; we later shared more than a few opportunities to share colorful Tom Lawrence anecdotes.)

I finally signed my employment agreement over dinner at the Restaurant Le Duc in Geneva on December 12th, 1980, the day of Apple's IPO. I recall inwardly grumbling I would not be getting pre-IPO shares. Factually correct and, as we'll later see, irrelevant.

Returning to Paris, I immediately get my first taste of Apple Is Doomed predictions[6]. The litany ran long. It started with the Apple][not running CP/M, the official operating system of the moment, before Microsoft DOS became popular; it continued with a display width limited to 40 characters and went on to the lack of 8" floppy drives — to say nothing of an upcoming company called Fortune Systems which would erase Apple from the face of the Earth… according to its founder, Gary Friedman.

But this isn't what I hear during visits to three Apple retailers: a ComputerLand franchise in Paris' Beaugrenelle shopping center; Jean-Louis Orsini, a small, technically versed independent computer dealer in Boulogne, an adjoining suburb; and Jean-Louis Cleenewerck, the head of prosperous Sivea, arguably Paris's best Apple shop. They are all happy with the product and say they'd welcome a more direct connection to Apple, better information, better tech support, and better access to spare parts for repairs. They aren't too unhappy with Sonotec, Apple's French distributor, but know the arrangement cannot last because Sonotec has little incentive for long-term investment in Apple.

I soon meet Georges Zimmeray, Sonotec founder and CEO. Georges was once an importer of hard wheat products from Morocco, as attested to by his SEMOULE (semolina) Telex subscriber name. Orthogonally interested in technology, he noticed a Californian company with an unusual name and reputation, flew to San Francisco, showed up unannounced in Cupertino and, checkbook in hand, offered to place a sizable order and be Apple's French distributor. Busy

---

[6] For almost 40 years, these predictions followed Apple at every turn. The Cassandras finally™ turned the volume down when, in 2020, the company's market capitalization passed $2 trillion.

developing the US market, Apple management accepted a 'light' contract with no long-term burden on either side.

I find Georges well aware of Apple's plan to start its own subsidiary in France and generally well-disposed, ready to accommodate an orderly transition if his help is properly compensated. Later, watching contentious transitions in other European countries, I came to see how exceptionally diplomatic Georges was; he made my job relatively easy.

A few weeks later, I hop on a plane to visit Apple's Cupertino HQ. That first visit turns out to be a revelation, although decidedly mixed.

I land late on a dark early February night and drive down the 101 freeway to my hotel, the Sunnyvale Hilton. Because they are overbooked, I end up in the last room available, the bridal suite — complete with shag carpet, four-poster bed, and mylar mirror on the ceiling.

The next morning finds me in the company boardroom with Steve Jobs sitting cross-legged on a credenza, picking his toes. He gives a funny look to my conservative pinstripe three-piece suit, asks what I am doing there, and offers unmemorable comments about Apple's future in Europe. That first conversation goes nowhere and concludes with a banal musing that the enemy is us.

I then go to my host, an HR person, and ask when and where the new employee training session will be held. No such thing. This isn't like the serious training and occasional vetting I lived through at HP and Data General. Never mind, having marveled at the elegant simplicity of Steve Wozniak's Apple][design while reading its technical manual on the plane — it reminds me of DG's Nova[7] design — I ask if I can

---

[7] https://en.wikipedia.org/wiki/Data_General_Nova

take a machine to my hotel room to train myself. This practice worked well in my past at HP, earning me respect inside the company, along with my joy at mastering complex products on my own.

Getting an Apple][proves a bit difficult and my embarrassed HR host lends me hers.

Back in the bridal suite, I experience a revelation: the power and simplicity of the VisiCalc spreadsheet program (we didn't say 'app' yet), one of Apple's best sellers.

VisiCalc holds a special resonance for me because I once tried to write a spreadsheet simulator. This was at HP when I thought of myself as an agile programmer. Freshly promoted sales manager of HP France's desktop computers business, I wanted to automate sales forecasts and reporting. To do so, I planned to simulate the green forms accountants used, called spreadsheets. The idea was to use the matrix function found on HP's time-shared BASIC system — but it ultimately led nowhere.

Here, on the Apple][screen, I see rows and columns of cells, each cell containing a formula referencing other cells, recalculated each time I hit the Return key. I sum columns and rows, assign growth percentages, see results, change assumptions, and see the new results right away. VisiCalc's design was elegantly simple. VisiCalc was built as an interpreter, working through a list of cells, and recalculating them when changes were made, and the user pressed the Return key. No fancy matrix manipulation. My VisiRevelation was in the bridal suite of the Sunnyvale Hilton, recapitulating the transformational experience of thousands of others.

Decades later, we users of Excel, Google Sheets, and Numbers find this banal. But it wasn't then — in businesses large and small, people loved VisiCalc.

The next day, still besotted with VisiCalc, I drive straight through a mandatory right turn and am pulled over. I do what we're taught to do in France: I get out of the car with my hands down and palms out, a gesture *Naked Ape* author Desmond Morris said would communicate open and peaceful intent. The cop hadn't read the book. Fortunately, seeing my three-piece pinstripe suit and befuddled mien, he takes pity on me, tells me I could have been shot, and explains the US protocol. It may have helped that he was audibly amused by the 16-year-old photo on my driver's license. He lets me go.

That pinstripe suit would make one last appearance in an East Palo Alto bar.

When setting up my first Cupertino trip, I contact an old acquaintance from my DG years; you'll hear more about him in Chapter *"Data General"*. An exceptionally gifted software engineer with a strong command of hardware intricacies, at that point he worked on Intel's promising next generation of processors, the iAPX 432[8]— one that was to prove too far ahead of its time. He invites me to his place for drinks but once I get there, instead of a beverage, offers to show me a particularly striking clip of Ridley Scott's Alien[9] after a line of cocaine. According to my host, this treat must have been what the director had in mind when making the movie. I decline and ask what happened to his going dry when we worked together in Paris. He shrugs, calls me a square and, instead, takes me to a bar he calls *pittoresque* in his broken French, a place called Boondocks on Willow Road. He says we can have a few beers and play billiards there.

---

[8] https://en.wikipedia.org/wiki/Intel_iAPX_432
[9] https://en.wikipedia.org/wiki/Alien_(film)

Picturesque it is, starting with a parking lot filled with big Harley Davidson and Indian motor bikes my friend calls hogs and a lonely BMW Motorrad that looks out of place. Then I walk into an American movie set. Let me explain. Like many French people who don't know the US, I thought biker bar movie scenes were carefully set-up caricatures, with bearded faces under big hats, beer bellies pushing garish t-shirts, loud arguments, Wille Nelson music... But this night I realize that the movie scenes of my memories were realistic depictions, and I was in the real thing. Later, replaying my first impression, I saw my error: American movies were made primarily for American audiences, and so couldn't veer too far from real bar scenes their audiences were familiar with.

Back to the moment, we have a problem; my pinstripe suit doesn't fit the scene and my host strains to explain why he, a respected regular, would bring in such a square. Still, I'm reluctantly served an insipid beer, but the contretemps shortens the night – and sends me to a Stanford Mall clothing store the following morning.

Later, in more comfortable surroundings, I have another revelation, this time in one of Apple's Sunnyvale warehouses on a street called Java or Bordeaux. I see, as expected, rows and rows of Apple][boxes. But there are also palettes of software programs Apple distributes. Software on palettes! Looking more closely, among other products, I see AppleWriter boxes. Coming as I do from Exxon Information Systems, which sells an expensive full-screen Vydec word processor for professionals such as attorneys and accountants, AppleWriter running on the Apple][tiny screen looks puny and amateurish. But I quickly count palettes, ask about weekly movements, and realize AppleWriter outsells not just Vydec but the entire 'professional' word processor industry.

There was more. Haunting the Cupertino office hallways, I find out the company publishes an Apple Magazine with, if my recollection is correct, a story by "Jonathan Livingston Seagull" author Richard Bach in one issue and a Ray Bradbury poem titled "Ode to the Quick Computer" in another. I only recall the poem's last verse: 'So, cowards, what are you afraid of...', which I would later use to gently poke fun at students in France's School of Government (ENA).

I had found a — but really *the* — computer company positioned at the crossroads between technology and liberal arts — a claim Steve Jobs would make many times once he reclaimed to the helm in 1997.

This, VisiCalc's elegant magic and warehouses full of software; together thrill me: joining Apple now feels even better than I thought when signing my employment agreement. I find Apple more exciting than good old HP, more promising than ambitious but unfinished Data General, and better aimed than lost-in-the-world-of-bits-and-bytes Exxon Information Systems.

To understand the source and strength of my sentiment, as explained in my life's first turning point, attending a Roman Catholic boarding school would build character for many.

Turning Point 3:
The Psychosocial Moratorium

Less than a year after my first love's death, another catastrophe— and an opportunity

One Spring morning, I saw my father marched out of our Paris apartment between two gendarmes, the result of accounting misdeeds. My divorced mother was at the other end of the world; I was left to my own devices.

For reasons not to be analyzed here, I saw my father's disappearance as an opportunity to become my own man instead of merely someone's son. I would work, make a living, and have my own life. This was the beginning of my psychosocial moratorium[10], where responsibility is suspended while the individual transitions from adolescence to adulthood.

But phrases such as "left to my own devices" and "become my own man" were more aspirational than factual. As I started pinballing around the job market, I found helping hands at each bounce. Friends, distant family members, and total strangers always materialized when the need for the next job arose.

France was in the midst of the "Glorious Thirties", the three decades following WWII, when the country was on the rebound. Jobs were plentiful, a blessing for someone as impatient and impervious to his inexperience as I was. In the four years that preceded my joining HP and entering the tech

---

[10] https://sites.google.com/site/motivationataglanceischool/p-theories/psychosocial-moratorium

Voie Royale, I went through a medley of jobs described here in no particular order.

I sold healthcare insurance to French kulaks[11]. In the early sixties healthcare coverage wasn't universal. Communist-infused French ideology didn't like independent, non-unionizable grocers, farmers, plumbers, and architects. I pitched them inexpensive, no-frills coverage for themselves and their families.

Here beginner's luck struck with great force. My sponsor at the insurance company, a family friend, pointed me to a prospect who had answered a direct-mail campaign. An architect with young children, he wanted and could afford full, expensive coverage which generated a sizable commission, enabling a vanity purchase: my first made-to-measure pinstripe suit. This was a world of hard, not always completely honest selling, and after a while I soured to its daily grind of turndowns by hard-nosed shopkeepers and artisans.

Fortunately, my deceased inamorata's father helped me study for and land a job as an information desk clerk at the *Gare Montparnasse* rail station. A demanding post, we clerks were judged on our ability to memorize schedules (no Web browsers back then) and decipher the ingenious codes in the French and European timetables. The detailed and cryptic tasks were invigorating, at least for me, with an unexpected benefit: A few times a week, a seemingly befuddled traveler would hand me a generous tip in exchange for a meticulously threaded itinerary that wound through Europe's rail networks. Five francs at the time (more than $10 in today's terms) provided lunch money for a couple days. Only later did I understand the nature of the

---

[11] https://en.wikipedia.org/wiki/Kulak

transaction: These inquirers weren't actually going anywhere; they just wanted to dream and thanked me for building them an itinerary in a pleasant fantasy world.

This was part-time so I also got a job as an office gofer in a small PR shop and where I had my shortest and most tragic job. I had barely counted the envelopes I was told to sort when a commotion took me to the boss's office, where I found him sprawled in his chair with pills spilling from open bottles on his desk. He wasn't breathing and had a weak and thready pulse. I, trained as a boy scout, attempted mouth-to-mouth resuscitation while other employees called "Police Secours" (a sort of Parisian 911). This ended horribly as the dying individual vomited his undigested lunch in my mouth, throwing me into uncontrollable tear-filled spasms. A neighbor doctor soon showed up and off-handedly pronounced the man dead. I washed my face and left for the next bistro, badly in need of a double shot of rum.

I joined a pharmaceutical company as a sales rep. We 'detailers', as the position is now called, visited doctors and hospitals to promote the company's antibiotics, among which was Rifamycin, a drug that completely cured tuberculosis (at the time, anyway; new drug-resistant forms have recently emerged in prison populations). This took me inside the French healthcare system where I saw abuses, cozy relations with drug companies, and the cynicism of doctors and even interns who quickly became accustomed to being wined and dined. As detailers, we thought that surely this corruption and decadence couldn't last.

I chauffeured the VP of Sales for Sud Aviation (a predecessor of Airbus), a tall gentleman of Polish extraction and regal bearing who favored Citroën DS saloons. In addition to developing what would much later become an obsession with smooth handling and braking, I saw how generously the staff at luxury restaurants treat their customers' chauffeurs, and learned how a good driver can make his boss' life easier: "Just keep circling the Printemps (department store), you'll see me at the curb when I'm done". Later, when I drove the family of a French impresario in stony, know-your-place silence, I realized how lucky I had been with the agreeable Sud Aviation VP.

I also did a stint as a door-to-door salesperson hawking office equipment. This was hardcore: No salary, no expense account, but fat commissions and invaluable training in the many facets of getting doors opened and objections parried and (gently) overruled. The training came in handy at a crucial juncture during my HP job interview, which I will cover in the next chapter.

I'm keeping for last what I later called my Anthony Bourdain period, a trip through the 'hospitality' business.

Early during this succession of jobs, a college classmate convinced her father to introduce me to the head of beverage operations at the Deauville casino conglomerate. That middle-aged gent started his career in the basement, pressing peaches for *champagne-pêche* cocktails, and slowly rose to the Big Job running multiple hospitality locations in the Barrière empire.

I lucked out. First, I showed up for the interview in the new, vested pinstripe suit I bought with my first insurance commissions. This helped me fit into that high-end place, and I got hired. Second, I was assigned to the casino's choicest location: the boardwalk Bar du Soleil. Finally, my boss, Jean-Claude Denos, was a fast, cocktail expert with an equally encyclopedic store of anecdotes on patrons, both shady and shiny.

I worked at the bar for the spring and summer seasons and had a grand time serving a mix of artists, jockeys (who routinely sweat off a couple pre-race pounds at the *hammam* nearby), and bookmakers (an illegal but tolerated profession in France). Customers were white-collar only; we had sneaky ways of turning away would-be blue-collar patrons.

Work regulations were happily ignored, which provided opportunity to make more money by working late afternoon gigs at the polo field or casino. That's when I met my first spouse, in a little town nearby. I was generously adopted by her family. The marriage only lasted ten years, but my gratitude lives on.

On the surface it all sounds good. But it didn't take long to discover the underside of the business. Shortly after I arrived an audit determined that theft was rampant. Shrinkage climbed as high as 40%; staffers stole entire hams and beef tongues (a sandwich delicacy), carefully wrapped and hidden at the bottom of a trash can. One waiter slept with one of the cashiers and used that relationship to perform a substantial volume of 'black' transactions. Almost everyone was preoccupied with finding cracks in the system to exploit.

When the summer ended, I was offered a place in the casino's croupier school. Instead, I wrecked my Citroën 2CV after a late-night gig. I fled to Paris where, with a recommendation from Monsieur Denos, I got a job as *maître d'* in a Montparnasse strip joint.

This colorful place had a gimmick: A comedian would break raw clay plates over his head, and then walk into the audience and break more plates over the heads of patrons. (The clay was soft and brittle, no harm done.) We then handed out more plates, often leading to a drunken pandemonium.

Now add female escorts to the mix. They were in the business of sweet-talking patrons into ordering more food and drink, while the waiters fattened the bill by removing and re-serving the same drinks and overcounting bottles. Sex was also part of the deal; I was to douse the lights in the alcove where an escort known for her digital prowess took care of business. I recall a group of coal merchants who regularly spent what today would be tens of thousands of dollars in one night. When another group, French Communist Party functionaries, left suitably drunk, we sang the Internationale as they filed out, stuffing tips in our white jackets' breast pockets.

From there, I worked as a waiter at a restaurant nearby, where I failed to whip a proper mayonnaise and saw customers (and staff) served almost-rotten food with lackadaisical abandon, while the chef and owner enjoyed themselves in the backroom.

Over three decades later I opened Kitchen Confidential on a flight to Osaka. The rightly mourned Anthony Bourdain describes the frenzy, the camaraderie, the cheating, the alcohol and more of life in the restaurant business. Before reading the book, I thought my experience had been a crazy fluke. But, no, Bourdain's book gave me a jolt of recognition: I had experienced the 'normal' underbelly of hospitality.

I'm glad I had those picaresque experiences, which gave me a lifelong appreciation for the hospitality trades as well as useful habits of observation. They gave me context for the enthusiastic embrace of my first job in the tech world.

Turning Point 4:
Joining Hewlett-Packard

I was lucky to join the company just as it started its ascent to the top of personal computing. This was the beginning of over 50 happy years in the amazing tech world.

"Great, we'll get back to you."

With encouraging smiles, my two bosses-to-be, HP France CEO Pierre Ardichvili and Sales Manager Gilles Bastien concluded the job interview.

I had a better idea:

"Why don't I step out in the hallway while you discuss your decision and then call me back in?"

The hard-core door-to-door sales experience I described in the previous chapter taught me to never take a dilatory answer. It paid off; my impudence was met with a pleased chuckle:

"Boy, you're salesy! Alright, call Gilles at 4 pm".

I did, and Gilles told me they'd take me "on approval", pending the result of an upcoming training session to start two weeks later in Geneva. "In the meantime, good luck with your next field sales trip. Call us when you get back."

During the interview my future bosses unfolded a brochure[12] describing the machine I was to launch in the French market.

12

http://archive.computerhistory.org/resources/text/HP/HP.9100A.1968.10264616
4.pdf

It was exciting, with a keyboard featuring programming primitives and esoterica such as hyperbolic trigonometric functions. I recall feeling more excited by the product than my interviewers were. In retrospect, my feelings must have been apparent, and probably swung the decision.

Two weeks later, I call as directed and am told "things have changed." The words sound ominous, but it turns out that the "on approval" clause of my hiring has been lifted for obscure reasons of confidentiality. "Come see us this coming Friday; we'll give you a plane ticket to Geneva and a cash advance for your expenses." I'm early for the meeting and wait for the appointed hour at a nearby bistro — where I spill chocolate on the lucky suit mentioned in the previous chapter.

The HP 9100A[13], the desktop machine I saw for the first time in Geneva, was an engineer's dream.

As I sat in the training room, I contemplated the convergence of opportunity and preparation (*a.k.a.* luck) which brought me to this tech industry trailhead. I felt I had finally found my path — and I was right!

After I escaped from the hospitality industry I landed in a Paris suburb just down the street from electronics giant Thomson-CSF. The local newsstands, which carried hard-to-find professional magazines such as *Electronique Industrielle,* were heavenly reading rooms for this long-time geek.

One day in May 1968, right after the student riots, I saw a tiny ad in *Le Monde* (France's equivalent to the *NY Times*). Hewlett-Packard was looking for a salesperson to introduce a 'revolutionary' product to the French market. Hewlett-who? I doubt *Le Monde* readers knew anything about the company

---

[13] https://en.wikipedia.org/wiki/Hewlett-Packard_9100A

at the time — but I did. Stories and ads in *Electronique Industrielle* had prepared me well.

My picaresque schooling didn't include résumé writing. Instead, I sent a two-page letter to HP explaining why they should hire me — and scored an interview. To this day, I marvel over my good luck and its repercussions on my family over the following fifty years.

I inhaled everything about the 9100A, its technology (which was quite strange in retrospect, especially its retroactively baroque two ROMs), its programming techniques and the applications it enabled. I started reading manuals in bed on Saturday mornings, a lifelong habit that served me well with colleagues and customers at HP and later.

My training wasn't limited to classroom sessions in Geneva, where HP had its European HQ. I was with the US product team when they introduced the product in London. There I met the inventor, Tom Osborne[14], along with Barney Oliver[15], HP's VP of Research.

That field tutorial helped me organize my first press conference at the Orly Hilton, where I met a few electronics industry journalists. It wasn't very successful. In my haste I left the 9100A at the office and had to delay proceedings. My exhaustive demonstration fell on deaf ears, largely because the product category wasn't sanctioned yet, and the phrase *personal computer* had no currency yet.

Successful press conference or not, it was time to go into the field and meet prospects. Kléber Beauvillain, a senior measuring instruments sales gentleman, took me under his wing as he visited some choice customers. These customers

---

[14] https://history-computer.com/hewlett-packard-9100a-complete-history-of-the-hp9100a/

[15] https://en.wikipedia.org/wiki/Bernard_M._Oliver

treated Beauvillain as a friend and were equally enthusiastic about the machine. They showed me ingenious techniques to overcome the 9100A's limitations with its mere 196 program steps and 16 storage registers. (Years later Beauvillain became the head of HP France, a position he held for close to twenty years, and in 1991 he helped secure our first investor at Be.)

I quickly assembled a sales team of like-minded techies who pitched the 9100A to engineers, mathematicians, and physicists across France. As part of my standard sales pitch, I disassembled the product and then put it back together as I lovingly described each component. (Years later, I used this fun and effective demo technique for product launches, including the Mac Portable introduction[16].)

Our prospective customers loved it, but sales weren't always easy. The French government's *Plan Calcul* drove public funds towards native computing technology and prevented state-subsidized entities from buying our products. But, as the saying goes, *impossible n'est pas Français:* French culture sees laws and regulations as mere challenges to creativity.

For example, the techies at Nord Aviation, an Airbus precursor, couldn't be seen using the forbidden machine, so they ordered a set of parts that, when re-assembled, looked remarkably like a 'genuine' 9100A. When I visited the happy rule-breakers they led me to a toilet, unlocked a mysterious door, and, laughing at their cleverness, showed me their 'homegrown' machine, sitting atop a commode filled with plaster (of Paris, naturally).

At the home office it didn't take long to see that HP France had developed its own libertine version of the gentle,

---

[16] https://www.youtube.com/watch?v=ZzIQdWKmPuE

civilized culture so accurately depicted by co-founder David Packard in his 1995 book *The HP Way*[17]. In those turbulent post-1968 times, when the law of the day was Forbidding Is Forbidden, HP France was what we now call a "den of fraternization". I must censor myself, as I did for the previous chapter. Still, I will recall the farcical situation in which a married woman, her lover, and her mother-in-law sat at adjacent desks in an open space office. Or how a crying spouse waved a pistol (unloaded, we later discovered) at the receptionist, demanding she return the husband she had stolen. Just one problem: It was the wrong receptionist. The intended target was enjoying a relaxing 'lunch break' in a nearby hotel.

I had my own caper. Desperate to land the job at HP, I had suggested that I had completed my military service when, in reality, I merely had a deferment. When pressed to return to the Army I used the pharmacological knowledge I had acquired as a drug 'detailer' to create symptoms which got me into the psychiatric ward of Paris' Val-de-Grâce military hospital. It was an interesting scene where inmates covertly looked at each other, trying to assess who was really crazy and who was merely trying to duck military obligations.

After a short time in a padded cell, I was free to walk around the hospital, where I received visits from my boss Gilles Bastien (silently impressed by my canard) and team members who brought me my mail. Thanks to a mock doctor's sticker on the windshield that fooled the guard, my wife delivered my Peugeot 504, allowing me to leave the grounds, attend an important trade show or visit a customer, and be back, saluted by the guard in time for the evening roll call.

---

[17] https://www.goodreads.com/book/show/1694071.The_HP_Way

After a while the docs decided the Army didn't need another troublemaker, recommended I seek therapy — which I did a few years later — and set me on my way. I was free — and not fired for my misdemeanor, one of many reasons for my gratitude to Gilles.

Just as I was getting out, HP's personal computing business exploded.

In 1970, for successors to the beloved 9100 machines, HP made the bold decision to use discrete logic to implement a 16-bit instruction borrowed from its minicomputer line (no 16-bit microprocessors back then) and proceeded to roll out a trio of industry-shaking devices. In 1971 the 9810 followed the key-per-function organization of the 9100, with important improvements such as plug-in modules and considerably more storage. In 1972 the 9820 surprised everyone with a full algebraic language, a powerful incitement to write clever, impenetrable one-line programs, a geek's delight. The 9830, which ran BASIC, came later the same year, sporting a better printer and magnetic storage. A hard disk was also on the horizon.

HP completely obliterated its competition by covering three distinct computing styles and price ranges.

But that wasn't all.

Soon after the 9100A introduction, Bill Hewlett insisted on a handheld version that would fit in an engineer's shirt pocket. The high-pressure development yielded the HP-35, a machine which surprised everyone, HP included, when it sold about ten times more than forecast. General Electric alone ordered 20,000 machines. You had to be there to experience the HP-35 shock. The next time I felt something similar was the 2007 iPhone launch.

HP promptly and hungrily set out to conquer the product category, creating an obscenely successful financial version, the $395 HP-80 (which is said to have contributed the majority of HP's profits in the difficult early '70s) and the programmable HP-65, with a magnetic stripe reader for external storage.

Offering both mobile computing and desktop machines in 1974, HP was the king of Personal Computing. The lovingly curated hpmuseum.org provides an epic list of HP machines.

Buoyed by my enthusiasm for the product and my sales team's good numbers, in late 1973 I was promoted to a sales management position at HP's European HQ in Meyrin, Switzerland. I came to regret that upward move. Unlike the colorful field life, I enjoyed in France, that bureaucratic position left me bored and distraught. The constant travel to other HP offices across Europe precipitated the end of an already problematic marriage.

Fortunately, an HP alumnus at the time put me in touch with a Data General VP looking for someone to turn around the company's French subsidiary. But before we move on to my Data General days, we should consider HP's fall from its personal computing pinnacle.

Musing: Why HP Fell

In the seventies HP dominated personal computing. Many attribute the company's slide into fragmented mediocrity to a series of bad board moves over the last two decades. But really, irreversible bad decisions took place much earlier.

[My debt to HP is immense: in 1968, they took a street urchin, taught him good business manners, inducted him in the wonderful world of personal computing, and pushed him up the management ladder. Finally, HP set me free to use my experience and credentials elsewhere in the computing industry. This chapter is the lament of a grateful but deeply disappointed fan and major beneficiary.]

Board scandals, bad CEO hiring decisions, and at least one momentously bad acquisition provide a ready-made explanation for Hewlett-Packard's decomposition. As an old-time (1968-1974) HP alumnus, I saw the trouble start much earlier.

Hewlett-Packard's decline began in 1999 when then-CEO Lew Platt, having spun off the measuring instruments business into Agilent, bucked tradition and recruited an outsider as his replacement. Carly Fiorina's six years at the helm were marred by unfulfilled promises and the catastrophic $25B acquisition of Compaq, a company wounded by low margins and its own disastrous acquisition of DEC for $9.6B.

Mark Hurd, Fiorina's successor, offered hope for a brighter future but promptly tripped over (never-confirmed) allegations of 'fraternization' and sophomoric expense account fiddling meant to cover up the supposed liaison.

In came Leo Apotheker, and out went less than a year later, thanks to the ill-considered Autonomy acquisition and rumors of his intention to spin off the PC business.

Meg Whitman replaced Apotheker in 2011 and tore the company apart. She 'liberated' the PC and printer business as HP Inc. and stayed on as head of HP Enterprise. After several quarters with uninspiring results — Bloomberg awarded her the 'top' spot on their list of underachieving CEOs for 2013 — she resigned in November 2017. (To be fair to Whitman and offer some perspective on such lists, Tim Cook once found himself ranked 13th 'underachiever' on that same list.)

Add in repeated incidents of board strife and an internecine spying scandal, and you get the full picture, from Fiorina to Whitman and on into second-class corporate status.

Or so goes the conventional Silicon Valley wisdom.

But HP's troubles were visible years before Carly Fiorina.

In November 1970 I took my first US trip, to the HP offices in Loveland, Colorado, where 'my' desktop computer products were designed and made. I arrived at night, and when jet lag woke me up, I opened the curtains to a mile-high Technicolor scene: impossibly exaggerated red earth, white clouds and blue skies, evoking John Wayne movies. But it was real, the majestic Colorado landscape. As an introduction to the US, it can't be beat.

I hopped in the "not too big" rental car (a mere Ford Galaxie) and drove to the office for a training session. I still remember how relaxed and hospitable people were, holding hands to say grace before a dinner of 28oz porterhouse steaks and weak coffee at The Rear Of The Steer (I'm not joking — that was an actual restaurant, dimly lit, as the custom of the day dictated). On a frigid Saturday afternoon, I saw my first college football game, complete with pompom girls (cheerleaders) and cannon fire to celebrate touchdowns. For a trip to a Las Vegas convention, a generous host took us aboard his own twin-engine plane and, after crossing the Rocky Mountains flew us inside the Grand Canyon, something forbidden today. I know now this was a very limited view of my future country, but the memories remain strong.

The HP I saw was the powerful, benevolent, prosperous technology company that started Silicon Valley before World War II. But the company was already torn between two cultures: On one side science lab engineers were designing and selling high-precision voltmeters, oscilloscopes, and Fourier analyzers; the other was the sexier computer business, propelled by the unrelenting progress of silicon-based logic. The tension bubbled up in petty ways, such as arguments over office decor. The instruments team believed in traditional plastic tile; the computer engineers wanted fancy, expensive carpet (again, I'm not joking).

Despite these differences everyone lived by HP's unwritten credo: We design products for the engineer at the next bench. This worked well for the instruments business, and also manifested in HP's new line of desktop and mobile computers. These were machines only an engineer could love — or even use. (The one exception was the outstanding financial HP-80, descendants of which, the HP-12C, we still see in offices today.)

True to the credo, HP's instruments and computers were more bought than sold, paid for by one's employer (invariably some sort of lab). HP had no empathy for individual users; it saw no profit in marketing to "mere mortals". According to one persistent and relevant industry joke, HP was so inept at marketing that if they sold sushi, they would position it as "a small ball of cold rice surmounted by a piece of dead fish".

An illustrative example: In 1977, HP created the visionary HP-01 calculator watch[18].

A technical tour de force, the HP-01 contained six IC chips for an equivalent of 38,000 transistors. (In comparison the 6502 microprocessor inside the contemporaneous Apple computer had fewer than 4,000 transistors.) But to HP a watch wasn't a computer, so it was only sold in high-end jewelry stores.

Another example: The brilliant HP-35 was only sold by direct mail. Why? The official explanation was not enough margin, but the real reason is that the company was repelled by and ignorant of the retail business.

At HP France, I found a way around the edict: I got a Paris slide-rule retailer to collect customers' mail-order forms in exchange for a measly commission gingerly authorized by my bosses. Orders flooded in, and at a trade show where actual selling was forbidden, we rented a small store outside the building and delivered machines to eager customers.

In the mid-70s things changed. Apple started in 1976 and Microsoft in 1975. They were parts of a movement fueled by affordable 8-bit microprocessors from MOS Technology, Motorola, and, most notably, Intel. Ohio Scientific, Victor, Commodore Tandy, and dozens of microcomputer companies sprouted in the late 70s. As noted on the Hewlett-Packard Wikipedia page, Steve Wozniak designed the Apple

---

[18] https://en.wikipedia.org/wiki/HP-01

I while employed by HP, and offered them the right of first refusal. Wozniak says HP turned him down five times.

Why would HP want a puny 8-bit home-brew machine? It already had a "personal computer," its ultra-sophisticated 9800 line, which culminated in the 16-bit 9830 with optional modules for terminal emulation and matrix computations, as well as compatibility with hard disks from the company's minicomputer line. It was clearly superior, but at $10K (at the time), it failed the true PC test: a machine you could lift with your arms, mind, and credit card.

We're left with a mystery: Up through the mid-70s, HP was unequaled in its creation of a world-class line of desktop and mobile computers, including a Because We Can smartwatch. Then nothing. Other computer companies caught on quickly: In 1976, Data General saw the tide turning and rushed their 16-bit microNOVA design down the street to their Sunnyvale factory. Despite its wealth of money, people, and industry connections; HP didn't see fit to design its own 16-bit microprocessor.

Why?

Outside tech, consider the world-class Volkswagen Golf, introduced in 1974 to replace the beloved but outdated Beetle — a perilous exercise. But the remarkable story unfolded as, year after year, VW kept improving the Golf without corrupting its identity: a sexy, modern small car — but not a boring appliance. VW transitioned but never lost the plot. But HP did.

Blinded by its loyalty to engineers and its infatuation with the ozone of the lab, HP couldn't see that it was about to be PacManned by 'inferior' products. To paraphrase Saint-Exupery, HP's heart couldn't see 'it', the future of desktop and mobile computers. They lost themselves pursuing bigger minicomputers, which were exterminated by Sun Microsystems, which itself later fell to servers built on PC components.

In modern business terms HP was disrupted[19], it 'overserved' a narrow market with capable products but neglected a much broader set of customers who made productive use of less capable but still serviceable devices. As disruption theory predicts, over time, market success gives 'inferior' products sufficient capability to muscle up and ultimately confine formerly dominant players to a niche or finish them off.

HP had the means, but one or more leaders must have lost the desire.

We now have the modest, stable $7B Agilent and its own $5B Keysignt spinoff, the hard-to-characterize $29B HP Enterprise 'solutions' business, and HP Inc, the $52B PC and printers' business. Such a mundane end for a revered titan. Yes, we see a succession of bad Board decisions, but the illness started decades prior.

---

[19] https://en.wikipedia.org/wiki/Disruptive_innovation

Turning Point 5:
Data General

Good manners don't always make for a complete real-world education. Data General provided a helpful counterpoint to HP's gentle culture.

Five years into HP, I got the itch to climb the corporate ladder. I saw an unlimited future at HP and even dreamed of getting a job in the Loveland lab (Colorado) where my beloved desktop computers were engineered.

Unfortunately for me, thanks to the good work of the team I led, my wish was granted. In 1973 I landed a Sales Management position at HP's European HQ in Geneva, Switzerland, in charge of southern Europe.

It was a valuable title on my résumé, with equally valuable travel opportunities and cultural exposure. However, moving from the daily stimuli of directing field sales in France and animated interactions with customers, to a staff position in Geneva left me feeling empty. The frequent travel helped end a marriage that hadn't aged well.

Alain Dasté, an HP alumnus, came to my rescue. Dasté introduced me to Barry Fidelman, VP Europe of a young minicomputer company called Data General (DG), looking for a new CEO to turn around DG's French subsidiary. Barry and I connected immediately, and I welcomed the opportunity to get back to Paris, to run a field operation once again, and to upgrade my résumé with a CEO entry.

Cultural differences came at me quickly. Initially the East Coast formality at DG was disconcerting: HP's "Jean-Louis" became "Mister Gassée" and, by virtue of his Ph.D., Jim Campbell, the HR VP, was formally addressed as Dr. Campbell. Nothing of the sort occurred with the numerous PhDs at HP, my former employer.

But the ostentatious stiffness was mere window dressing.

At one of my first meetings in Westboro, MA, a senior executive proudly demonstrated his latest *objet d'art*: A new LED watch that, with the push of a button, displayed a clear invitation to carnal congress. In another staff meeting, to big guffaws, the same exec circulated nude pictures of an engineer's girlfriend posing on stout Control Data disk units.

A quick tour of DG's nearby manufacturing plant accentuated my sense of joining a different culture. An HP plant was a calm, clean and well-lit place; DG's was more like a forge, darker and messier, with people scurrying around without much sense of harmonious workflow. In another building I saw a high-temperature oven where the company cooked its ferrite memory cores; I couldn't understand why my new employer made something that already was a commodity product.

I met Ed de Castro, the company founder, who listened to my questions and suggestions for market segments with barely polite indifference. I later learned this had nothing to do with me — he was often lost in his own thoughts and didn't like suggestions.

More positively, I got kudos for my then-shaved skull, proof I was, in another burst of unprintable words, deranged enough to be thrown into the snake pit that DG France had become. He had mistaken an impulsive decision made during dark days at the end of my HP career and marriage for renegade bravura.

Ed was wrong about what my bare scalp meant, but right about the organization I inherited. My predecessor did not hold US businesses in high esteem. He had switched the local version of the black and orange DG logo to a blue-white-red combination echoing the French flag. This was part of a lame attempt to placate prospects and government officials concerned with protecting French suppliers from 'unfair' American competition. When US management objected to the distasteful anti-Americanism, he insisted, "No, no, trust me, this is how we must do things in France."

With my predecessor gone, I quickly put normal business practices into place, and the local staff seemed relieved.

At this point I discovered cash. Allow me to explain.

HP delivered working products, customers paid, and that was the end of the story. There we never had to worry about cash. Things were different at DG where products, often with advanced feature sets, rarely worked out of the box and needed finishing in the field — if we could get the needed parts and data. Even when we finally put together a working system, customers became reluctant to pay. Our local operation struggled to make ends meet.

There were two positive outcomes.

First, my boss, the aforementioned Barry Fidelman, saw how my comfy HP years had failed to teach me about hard business realities and promptly sent me to INSEAD, the noted International French business school. That remedial education still serves me when waving my BS detector over business plans and SEC filings.

Second, I learned from the savviest salespeople in tech. Wisely, DG paid its sales force good commissions, but with a down-to-earth twist: No payment on bookings or shipments, only on collections. Salespeople quietly listened to marketing's *missi dominici* preaching to the choir. They could recite religiously all the functions and features of the latest and greatest DG products. In the field they limited themselves to system configurations they knew customers would pay for, to produce commissions.

I liked DG's products, both hardware and software. These included Nova minis, system software, and the Eclipse computer — whose development was lovingly chronicled by Tracy Kidder in *The Soul of a New Machine*. As a self-educated geek, I particularly appreciated the technical manuals, each with an introductory "Concepts and Facilities" section to provide a cogent orientation into the product's *raison d'être* and main components. And then there was the biweekly delivery of technical, marketing, and sales documentation called MAPS. (I'd love to see another prosperous ex-employer of mine, Apple, follow the MAPS and "Concepts and Facilities" example. On a more recent experience, I'd like to see my favorite German automaker, of excellent but technologically overcomplicated vehicles, do the same.)

I had the opportunity to meet one of the MAPS authors, Ed Zander. He had a full head of jet-black hair back then, but he already had the animated speech and bursting-at-the-seams briefcase he would later be known as an executive at Sun and Motorola (where I met him again to discuss the sale of Palm Source).

After I performed a first round of clean-up work at DG France, my interest in the inner workings and finer details of technical products led to a move to DG's European HQ in Paris, where I was put in charge of Product Marketing. Thanks to my HP Europe past, I was also put in charge of Southern Europe and Middle East distribution.

This began a period of travel into Iran and Israel when you could still travel between the two countries (before the 1979 Iran revolution). I got my first taste of El Al (the Israeli airline) security. No x-rays — they stepped back and made *you* open and search your own luggage under their direction. If anything exploded it would be your hands, not theirs. And each flight featured a couple 'civilians', suntanned and in excellent shape, security officers in mufti. At this time in the late seventies, life in Tel Aviv and Jerusalem was reasonably peaceful, and on Friday nights miscreant Jews who wanted to eat, and drink were welcome on Jerusalem's East Side. Data General was welcome in Israel, and with our local distributor we sold minis to military and civilian tech organizations. In Egypt, another pleasant environment, Egyptian Army colonels considered their opposite numbers in Israel esteemed colleagues, and considered buying our Nova computers simply because the Israeli Army did.

In Kuwait our distributor, who also represented Cartier and L'Oréal, introduced us to a DG customer calling itself Kuwait Danish Dairy (KDD). Kuwait was hardly known for its green pastures, but KDD reconstituted milk from powder imported from Denmark, and then made all sorts of well-liked dairy (and improbably tasty — I tried the ice cream) products peddled in shops and pushcarts around Kuwait City. Our distributor even offered to sell me his Phantom V Rolls for $10,000 because his mother didn't like it anymore. I was tempted but failed to corral a road-trip companion to drive it back to Europe. I sometimes still fantasize about the missed experience.

Our trips to Saudi Arabia were almost as pleasant. When we landed in Ryadh the passport officer didn't want to let Barry Fidelman, a Jew, into the country. Our distributor wouldn't take that no, launching into a tirade invoking the Kingdom's

Constitution that, in theory, offered equal protection to The People Of The Book (Abraham's children), probably also mentioning connections and money instruments beyond my hearing, and Barry was let in. Dinners were held inside closed courtyards, sitting on divans, drinking fine spirits, and peacefully arguing about DG's distribution business's finer technical and financial points.

After the Ryadh visit we were to fly to Jeddah. Boarding the plane, we sat next to a robed gentleman holding a blinded falcon on his fist, feeding it raw chicken on a towel by his feet. It took a while for the flight captain to accept the arrangement, but eventually we had an uneventful journey.

Unfortunately, the easy and colorful journeys to Middle Eastern countries did not last, nor did my involvement with DG after Fidelman, a man of unusual intelligence and deft business touch, went back to the states. Barry was replaced by an ex-GE executive whose different style and substance made me susceptible to the siren songs of headhunters.

After six successful years at HP, my DG shock proved I still had a lot to learn about the real world of business, the sharp elbows, the flexibility of 'facts', and how to navigate markets in agitated countries. My five years of real-world education at DG were invaluable over the following decades.

Before we see how, at Exxon Office Systems, I learned how culture eats strategy — and billions — for breakfast, we need to discuss another turning point, of a more personal nature.

Turning Point 6:
Meeting Brigitte

After a failed first marriage, meeting and marrying Brigitte permanently changed my life for good. It might sound trite and formulaic but read on.

Between Brigitte and I, things didn't start well. In 1976, she answered a recruitment ad offering an executive assistant position for the Sales Manager at Data General France. She interviewed well. The Sales Manager and I (I was DG France's CEO then) decided to offer her the job. When she saw the salary we offered, she protested it was lower than the desired compensation stated in her letter applying for the job. We demurred. She said she needed to think about it. "Sure, while we get coffee, why don't you use a conference room with a telephone and tell us your answer when ready." She accepted the offer, grudgingly.

Later, after I went around open office desks and deposited copies of a magazine article criticizing some government policy for being anti-business, she and a newly made friend of hers came to me and demanded explanations for that overt act of propaganda. I didn't dwell on the fact that this was a private business or that I was the boss and instead, with a straight face I expressed my appreciation for their willingness to share their thoughts.

I liked her but, having seen the trouble fraternization causes, kept a safe distance. Fortunately, I moved to the European HQ on *rue de Courcelles* in downtown Paris. Since I was no longer her boss and there was a new CEO at DG France, I felt freer to approach her, carefully. My ruse, which she easily saw through, was to invite her to play squash. I picked up that sport thanks to the DG France Sales Manager, who took me to INRIA[20], at the time housed in post-WWII US military buildings which offered two squash courts — and cold showers. Brigitte knew of my weekly squash dates and professed little surprise when I invited her to such a healthy activity. This proved to be a one-time event: she was a smoker at the time; once she got painfully winded, Brigitte pronounced that game unfit for her. During drinks after the game, with post-exertion endorphins smoothing the mood, Brigitte realized I wasn't quite the ogre she'd seen at the office and accepted a movie date for the following week.

We were on our way to over four decades, years marked by three sets of signal events: raising a family; Brigitte saving my life and soon thereafter our relationship; and the commencement of her third career (still ongoing) designing, building, and selling high-end houses around Palo Alto and Atherton.

Our multi-generational family — three bicultural adults successfully embedded in the Bay Area tech world and two grandchildren who graciously let us adore them — is happily conventional.

As I write about these years, my chest rises as my heart sings memories which could fill their own book.

For instance, I recall a bedtime moment. I used to read our children books like Selma Lagerlöf's *Saga of Nils Holgersson,* one my mother read, or a mythology *précis*. One night I read an episode about Aeneas going down into the underworld and meeting his father. Our son complimented me: "You read it really well." He meant better than my sometimes-tired diction

---

[20] https://www.inria.fr/en

at the end of the day; he had sensed my emotion. My father had just died of liver cancer, and I arrived at his bedside too late to say goodbye.

As I watched our children grow, I achieved a happy realization: they were wiser than us grown-ups. Not yet encumbered by adult passions and debts, they told it as they saw it. As they were sitting on the other side of the kitchen counter, an insight came from the back-and-forth reflections of love. From my childhood, I had no memories of tender moments of loving words from my parents. Now, basking in my love for them, I had a fleeting experience of what could have been — and gained an even deeper joy at being a father.

Things weren't always so heavenly. I was very busy during parts of my life, especially the Cupertino Apple years. My agitation, my overactive temper, and a thin skin (made only thinner by cultural misunderstandings) sometimes darkened the days. But I had hope; it was sustained by what an old healer once told me years before I met Brigitte: Children are the first beneficiaries of therapy. It turned out he was right — and that our children were forgiving and resilient.

As the little ones grew, we took them on long-distance trips. Some entailed riding up and down I5 to Canada with other family members and friends in caravans of a Suburban and a passenger van. Other jaunts took the five of us to Saint Petersburg, Vietnam, and Cambodia. And, thanks to the knowledge acquired in many Japan trips for Apple, especially under John Moon's[21] expert and paradoxical guidance, I proudly managed to guide our journey to Kyoto and Tokyo. And road trips in France when the youngest member of the brood seemingly completely ignored *châteaux* and museums

---

[21] https://en.wikipedia.org/wiki/John_P._Moon

— but I've since heard her brag to adult friends about all these interesting places.

Yes, my heart sings, and I could go on, but we need to jump to Brigitte's 1993 intervention.

After leaving Apple I decided I was out of shape and hired a personal trainer, who came to our house three times a week. It worked: I lost and gained inches in the right places, and soon fell into the mirage of being stronger in my near fifties than in my twenties. I redoubled efforts, over-exercised and, about an hour after one session, experienced disquieting symptoms: loss of speech (aphasia) and a paralyzed right arm. I improved after a while and thought the indisposition would pass. Brigitte found me resting, asked the right questions, looked up symptoms in an encyclopedia (no web browsers yet!), and dragged me to the emergency room at the Stanford Hospital. There doctors diagnosed a stroke and saw an ultrasound image of a damaged carotid artery and clots. I recall seeing the image and hearing how the clots could move up to my brain and do a lot of damage: I could be dead — or worse. I got what 'worse' meant: vegetable. Neurosurgeons quickly got a more precise picture of the damage, and then repaired my left carotid artery. While I was in the ICU Brigitte insisted on staying in the waiting room at the door. Kind nurses took pity and gave her blankets so she could sleep there. I only learned later of her vigil and can't mention it without becoming emotional. After less than a week in the hospital, I was well enough to be discharged — wearing tiny metal clamps in addition to stitches on the left side of my neck. Brigitte knew I was much better when, being asked what I wanted to do, I requested a trip to the Palo Alto Fry's store, where I surprised acquaintances with my disheveled look and Frankenstein neck.

Once the Being Alive Again feeling wore out, I asked Meaning of Life and What Am I Doing Here questions. That's when

Brigitte's second life-saving intervention took place, saving our marriage this time.

After that frightening and life-threatening event she didn't want to go back to the old ways of my alternating bursts of anger and periods of gloomy silence. There was no ultimatum, and no need. It was clear she'd rather be close to her family in France, with our children, than in a culture that wasn't hers in a joyless marriage. Having experienced the painful, albeit very civilized, ending of my first marriage, and agreeing our relationship needed help, I sought advice. A kind local sage directed me to a couple's therapist, with whom we began work that healed our connection.

At first, I felt like a difficult dog taken to the veterinarian to be declawed. But our therapist, Hilda, painted a different picture. As she referred to me in kinder terms and, I bathed in her gentle but firm aura and soon felt differently about the process and myself. She also taught us how to have conversations at a privileged, private time of the day, a process that soon enough made the weather sunnier. When we asked her why the mood changed so quickly — and as it turned out so durably, she replied that there was a solid foundation of love between us; all we needed was to sweep out the layers of debris that masked it. As I write this, I realize this sounds like the kind of psychobabble I used to mock; this is often the problem with the ineffable quality of a therapeutic experience. We both came from broken families, and little by little, sometimes in big strides, we left behind or took better ownership of our respective baggage. In the beginning, we saw Hilda once a week, then every two weeks, and took to basking before her while we gushed about our children and moments together. After eight years she kindly fired us, saying she was pleased with our work together, that it made her feel good about her profession, but we needed to move on untethered.

Five years later, observing the prescribed social delay, we invited her to lunch at Berkeley's Chez Panisse and later to a showing of Brigitte's paintings at the French Consulate in San Francisco. There she was quickly surrounded by Brigitte's female friends, who all wanted to meet the magical Hilda they had so much heard about. When our children felt we were about to start an argument at home, they jokingly threatened to call Hilda. She is no longer with us, but her memory and our gratitude live on.

There was an added benefit from our work with Hilda.

Returning to 1985, when we discussed moving to California, Brigitte had mixed feelings about leaving her family behind. I tried to reassure her this would be for a few years, two or three, no more. For this I relied on my HP France memories: there promising individuals were sent on assignment to one of numerous HP divisions in the US; after a few years, they came back with an anointment that got them promoted to higher responsibilities. Brigitte wasn't convinced. She went along but didn't stop yearning to return to the Old Country. She insisted on sending our children (two born in France and a third in Palo Alto) to bilingual schools and took the entire brood back to France for Christmas and long summer vacations.

Then, after our relationship healed, she stopped yearning so much for a return to France. I noticed and asked why. It had not happened in a 'decided' fashion, she said. She wasn't sure why, but it gradually emerged that, because we felt closer and safer together, her need for her original family's bosom was less strong.

A few years later, after a successful voyage into abstract painting that spread into family and friends' homes, Brigitte started her third career: home designer and builder.

It had started with a simpler remodeling of a home in a Paris suburb. Then, for a largish (by French standards) house we bought in Palo Alto; we eviscerated the original building; reducing it to its bare bones and rebuilding a modified, modernized, and extended version. Brigitte managed the project with help from a high-end (expensive) San Francisco architecture firm because I was busy with my job at Apple. Initially we were horrified to see the process double the original price of the house (we later discovered this wasn't unusual). But when we first woke up in our new home, hearing the morning train in the distance, everything was forgiven: the redos, the change orders, the cost overruns. We were to live fifteen years in that house, and only sold it after the children went away to start their own lives.

Our eldest child, Paul, now in his early forties, is a people person. He has a gift for starting and maintaining decades-long friendships. He bloomed into a successful coach/consultant for startups. Engineers love to engineer but often lack a sales gene. Paul helps them build a sales process that connects customer pockets to company coffers.

Next, in her late thirties, Sophie is married with two children who let us adore them. During childhood she heard me grumbling against the consumer-type MBAs inside Apple's Marketing organization. I complained about their vacuous attempts to shoehorn their expensively acquired Case Method learnings into the company's unique situation. Perhaps to tweak me, Sophie got just such an east coast MBA, and now works for a Valley giant.

Our youngest child, Marie, in her mid-thirties, also heard my rants and promptly got an MBA (on the west coast this time) and built a successful career at a succession of smaller companies. Marie is known in her family as Camp Director for her repeatedly proven gift for organizing vacations.

Brigitte and I are very proud of them. We bask in their tolerance, adult affection, and wisdom.

Having moved to a smaller house, still in Palo Alto, we promptly rebuilt it. Here Brigitte took matters into her own hands, picked an architect and contractor, and learned to deal with Palo Alto's famously prickly and slow Building Department. Dealing with a 'historic' house built in 1903 came with surprises, but the end result is a pleasant abode we sold to friends a few years later.

The virus was firmly established.

In 2007 Brigitte bought her first 'teardown', planning to destroy it to make room for new construction. There she drove the whole process. Because we had accumulated a few assets, she didn't have to deal with construction loans, she led the exterior and interior design processes, managed the general contractor and subcontractors, and finally led the sales process with a realtor of her choice. Her first design came on the market in 2009, right after the real estate market collapse. She only made a meager profit for all her pain and skills, but she had honed the formula that still undergirds her work more than a decade later. People who don't know Brigitte too well sometimes politely venture to guess, "so, your spouse is into interior design."

No. Pulling out a narrative honed by many such conversations, I use an anaphora to explain how she is an A-to-Z real estate developer.

Once Brigitte has found a teardown she tortures her realtor to get the property at an acceptable price; next she tortures the architect to get the design she wants; she then tortures suppliers of materials, appliances and finishes to achieve the effect she has in mind; she next tortures the contractor to build the house on schedule and budget (when the planets align); she finances the whole project and, when done, tortures her realtor again to get a reasonable price for her work. I use the word 'torture' with only mild license, given the demanding path of such projects, from the time a suitably derelict property comes to market to the moment, typically 18 to 24 months later, when the buyer's funds finally find their righteous destinations.

My role in this is to be an admiring supporter, occasional adviser, always honest without deleterious flattery, and sometimes chauffeur. Gender stereotypes are what they still are; people working on the construction site sometimes talk over her head and address me as the boss. "No, I'm her chauffeur," or more often: *"No. Ella es la dueña, yo soy el chofer."*

As one easily imagines, our adult children greatly admire the mother who raised them and has been making good money building elegant houses in Palo Alto and Atherton for over a decade. For a confined Christmas, we kept gift-giving much simpler than in previous years. A small secret lottery matched giver/recipient couples with a low-dollar limit, and off we went to find inexpensive, creative gifts. Our middle child and her husband came up with a frame containing pictures of ten houses Brigitte built, surrounding a small map of Palo Alto with dots marking their locations.

We all marvel at my luck meeting Brigitte over forty-five years ago.

Turning Point 7:
The Exxon Delusion

Exxon's failed foray into the world of information systems proves once more that culture devours everything, to the tune of about $4B. How did the supposedly wise Boston Consulting Group not warn Exxon about incompatible cultures?

Late in 1978, Data General's VP Europe Barry Fidelman returned to the Westboro, MA mothership and a new boss descended from the sky — more accurately from General Electric. The company descended concomitantly, from Barry's thoughtful finesse to a more regimented drill management style. It was time for me to make a change.

After briefly fantasizing about a return to HP's warm embrace I looked at other opportunities. One of the two companies that presented themselves as needing a turnaround was too clearly behind the times, with aging mechanical printing technology. The other, an enterprise making air-filled packing material, felt too small and commoditized. It is still around, now living off e-commerce's prodigious consumption of packaging material — I think of it, without nostalgia, when I open an Amazon package.

Then, I was contacted by a headhunter— pardon, an executive recruiter. As you may have experienced, recruiters and their clients generally have no imagination; they function like Central Casting: Send us a 6-foot caucasian male, wide shoulders, lantern jaw, blue eyes.

I have none of these attributes, but I became known as a turnaround specialist after my adventures at Data General France, and that's exactly what the client, Exxon, was looking for.

After a short stroll along *rue de Courcelles*— the recruiter's office was practically next door to DG Paris — I was greeted by a smiling gent who proudly offered glossy press cuttings, including a Fortune Magazine article praising Exxon's visionary plan for a post-petroleum future. I was also treated to a set of slides from the Boston Consulting Group (BCG), the management consulting company Exxon hired after two destabilizing and disorienting Middle East supply crises. I don't know what psychedelic mushrooms or toad exudate the BCG shamans used on the Exxon execs, but it worked: Exxon's mantra became "Information Is the Oil of the 21st Century." Exxon Information Systems was born to lead Exxon into the post-oil future.

As part of the transformation Exxon wanted a more tech-savvy CEO for their French Information Systems affiliate. This sounded like a good opportunity to put my turnaround experience to work in a promising, forward-looking, well-financed context.

Sales and Customer Service staff seemed happy when I arrived at the office tower on the eastern edge of Paris. Months of uncertainty were dispelled — they were keeping their jobs.

The reception was frostier on the finance side of the house, where I could not understand the head of Finance's bizarre objections to my conventional requests. I initially felt sheepish and blamed my own inchoate financial education for the constant misunderstandings (you may recall my remedial finance instruction at the

INSEAD business school early in my DG years). Still, with some advice and confidence-boosting from an older friend, I seized the opportunity to deploy a theory I developed at DG:

"If I don't understand what you're saying, I can't possibly be the idiot. Either you don't know what you're talking about, or you're hiding something. Would you like to go back to your desk and come back when you're ready to make sense?"

It sounds arrogant, and it's only justified in the individual's area of responsibility — it would be unfair to ask a software engineer to explain how to account for goodwill impairment — but it works. An audit revealed irregularities. The gent had something to hide. I promptly brought in a replacement from my former employer (a friend who would follow me to Apple and be later taken from us by smoking-related cancer), and the misunderstandings disappeared.

Image #1

On paper the Exxon Information Systems conglomerate looked like a behemoth-in-waiting, circling the bits-and-bytes market from several angles, amply financed by the rich oil giant.

Undoubtedly, information is the oil of the 21st century so how is it that today's Exxon Mobil (XOM on stock charts) is only worth $476B versus more than $2T for Apple, more than $1T for Google, Amazon, and Microsoft, $470B for Meta/Facebook, and so on?

It wasn't for lack of money and effort. Soon after the BCG revelation, Exxon put about $2B on the table — that was real money before Uber and Snapchat billions — to buy Vydec, a word processor company; Qwip, a fax machine; Qyx, a smart typewriter; Intecom, a telephone switch, *a.k.a* PBX; and Zilog, a microprocessor company started by the illustrious and outspoken scientist Federico Faggin, whom I later befriended during my Logitech years.

These weren't toy technologies. Zilog created the well-respected Z80 microprocessor. The Vydec word processor system had an elegant, full-page vector display, much more legible (and pleasant) than the coarse dot-matrix display used by competitors. Law firms and other high-volume document producers appreciated Vydec's polish and professional features.

It was an exciting time, but I soon saw Exxon for what it was: an out-of-its-depth organization with no feel for the world of bits and bytes, hardware and software engineers, or end-user sales and marketing (see Image #1). For Exxon leadership these concepts were just that: ideas without underlying reality. It was impossible for them to make the right choices.

For example, Exxon execs felt they were being told tall tales by shifty entrepreneurs, and sometimes they were, but the execs couldn't judge. Nervous Exxon accountants overrode local management's authority all over Europe, demanding written advance justifications for the smallest office expenses, because they didn't know what was essential vs. petty office graft. It was an unbridgeable cultural obstacle; BCG had convinced the company to perform a dance for which it couldn't find the rhythm.

Sitting in my office on the 17th floor, after months of non-communication between the disparate acquisitions, I saw the negative power of incompatible cultures. Culture isn't a set of rational dicta. It works below our consciousness — processing, filtering, and labeling raw data before passing the result to our 'waking' selves. That's how we end up with Obvious Truths, that's how we get to the powerful and destructive It's How We Do Things Here.

I jump ship and look for my next assignment, and I know where I want to land. After visiting Loveland, CO, during my HP days, I long to return to the US.

Exxon would spend another $2B trying to make Exxon Information Systems work, a total of $4B wasted dollars. After years of failure, it finally threw in the towel, returning to the business it knew in its head and its gut.

Right idea, wrong culture — it's obvious in retrospect. I can forgive myself for not seeing it as a rookie executive. But how could the all-seeing, all-knowing Boston Consulting Group take money for that plan, when they must have known the culture was incompatible, no matter how good the idea?

Having reviewed my years before Apple, we can now return to the story we left at the end of chapter *"Turning Point 1"*.

VIDI:
THE APPLE YEARS

FRENCH DAYS

Setting up Apple France

Returning from our time travel excursion, we briefly stop in Cupertino before returning to France.

At the end of Chapter *"Turning Point 2: The Apple Impact"*, to understand my enthusiasm for my newfound employer, Apple, we went back to the fifties, to my childhood years and the birth of my passion for electronics. It's now time to go back to February 1981 and my first visit to Cupertino.

To complete my first exploration of the Apple mothership, I visited Apple dealerships on El Camino Real[22]. I got a friendly but disconcerting welcome: "Ah, you're from Apple Europe. This is great. But we never see anyone from Cupertino." That didn't happen just once, but each time I walked into an Apple retailer doing business a mere three or four miles from company HQ.

Back at the base, I went directly to the permed (yes, this was the day's male fashion for ostensibly straight individuals) VP of Sales and asked why he and his people never paid even courtesy visits to the nearby flock of faithful Apple dealers. After silently letting me know that neither my poorly concealed surprise at his hairdo, nor my question, were appreciated, the gent condescended to explain that Apple used manufacturers' representatives. These reps were independent firms, agents of sorts. They contracted with several manufacturers, visited their distribution networks,

---

[22] A historic North-South artery dating back to the times when California was a Spanish colony. https://en.wikipedia.org/wiki/El_Camino_Real_(California)

and performed various tasks — looking at sales, inventories and promotional activities, and pushing them for more orders. I later learned this was inherited from Intel and other semiconductor companies many Apple execs came from[23]. It worked for chips. But having sold desktop computers at HP, I doubted this would work for a technically and intellectually richer product such as the Apple][, or its peripherals and software.

These visits to local retailers and my conversation with the head of Sales had a deep influence on the way we managed our relationship with French Apple retailers,

The last notable event from my first Cupertino trip: my encounter with *Le Chat Mauve* (The Purple Cat) founders Philippe Chaillat and Didier Chaligné in a corridor of an Apple building known as Bandley 3 (which I would inhabit four years later). They were looking to sell their invention, a card plugging into the Apple][bus that would output to the standard Peritel connector in the back of France's SECAM consumer TV sets. The built-in Apple video signal was the mediocre NTSC[24] signal of US TVs. The *Le Chat Mauve* card outputted a gorgeous color signal that looked terrific on Sony Trinitron monitors, and more importantly on standard TVs sold in France. I was unhappy with the accepted but poor quality of Apple][output on standard US TVs, so I sensed an opportunity, and told them to give up on Cupertino people focused on the US market, and to instead come see me in Paris in a few weeks.

Back in France I started putting the pieces together for an organization whose main task would be to grow and support

---

[23] Mike Markkula, Apple's first investor, and second CEO came from Fairchild and Intel https://en.wikipedia.org/wiki/Mike_Markkula

[24] Nicknamed Never The Same Color

a network of retailers. I needed a team and facilities. Fortunately, my luck held.

When I signed my employment agreement, I thought, 'If I can hire these two guys, I'm saved.' The two individuals I wanted were Michel Delong, a serious logistics pro I loved working with (he sometimes said 'against') at HP France, and Gilles Mouchonnet, a just as serious finance guru I met at Data General and recruited to Exxon Office Systems. I animatedly described my Cupertino visit and got them to sign up immediately.

To start with, I needed a Sales Manager. After a few unsuccessful interviews I was fortunate to hire Jean Calmon, an ex-IBMer who blended his former employer's professional approach with the flexibility required to run an independent retail organization. I still see Jean, now retired, painting maritime landscapes. He penned a personal and family memoir titled *Ce que mon ombre m'a dit* (*What My Shadow Told Me*).

Sadly, years later, tobacco-related cancer took Michel and Gilles away.

Dealing with French authorities presented an unusual challenge. The problem was that the French computer industry needed protection from foreign ogres. Under French regulations, foreign investment controllers denied Apple permission to establish a US-owned French subsidiary. That didn't faze Apple's legal advisors at Ernst & Young, who found a way around the ruling. They put together a local structure owned by a French citizen, yours truly, with the smallest capital investment allowed for a company structure known as SARL (a Limited Liability Company) for 20,000 Francs (about $5K these days). No problem, I wrote the check and promptly became the owner of a company called Seedrin, the last letters of my name and of the street (Emile

Landrin) I lived on. This was perfectly, albeit perversely, above board. There was no side letter or under-the-table arrangement required because of what follows. Seedrin's role was unusual and ingenious. It used one of Europe's Common Market regulations originally designed to import produce. We at Seedrin were 'commissionaires', taking orders to benefit an Apple distribution company based in Zeist, Holland. We managed the order and merchandise flow without having title or receiving payment for it. We got a commission for our services: neither too low nor too high, but enough to pay Seedrin's operating expenses. All legal and outside the purview of France's investment regulators. A few retailers expressed surprise when receiving merchandise from and sending funds to a Dutch entity. When the reason was explained, these fiercely independent businesspeople loved the arrangement.

Next, we needed offices, a warehouse to accommodate merchandise flows and a service department. Louis Calvarin first headed the latter, and then Maurice Lenoir, two individuals I still enjoy seeing. Louis is now a Breton farmer near Brest, Maurice lives near Amboise and revealed himself to be a polymath: politics and economics analyst with a fine pen, horticulturist, and also oenologist to whom I owe many years of exploring fine French wines, mostly but not exclusively from his Loire region.

Talking about serendipity, I wanted our base to be in Orsay's *Courtaboeuf* industrial development, about 15 miles from Paris. I knew the area well as my old HP France haunt, and I thought it would make life easy for Michel Delong and the colleagues I hoped he'd bring with him. Well connected to the freeway network but not too close to the big city, we could afford the rent. We looked and quickly found a building abandoned by its owner, pharma company Choay. Their

CEO didn't quite understand how a tiny company like Seedrin could afford to rent twenty thousand square feet of office and twice that of warehouse in the back. I requested a meeting and explained our scheme and how it was designed to run around government apparatchiks. He was no lover of regulators and pronounced himself amused and admiring of our advisers' work. We promptly shook hands and got free office furniture in the bargain.

Today such arrangements wouldn't be possible. The Apple of those days was highly flexible; it is now a very strait-laced organization, understandably concerned with strictly adhering to local rules and regulations. As an example, me nominally owning Seedrin while acting like an Apple subsidiary would be unthinkable. And renting an office and warehouse complex to a 'front' company would be shunned by 21st century directors and auditors. Small, not always taken seriously, we 1981 minnows slipped through the nets of inattentive regulators.

We had a management team, an office and logistics facilities, so we could set to work.

Sales and Positioning Strategies

Propelled by instinct rather than strategic vision, we stumbled on a virtuous cycle of sales tactics, logistics performance and financial management.

We started by hiring technically competent salespeople to discuss the ins and outs of Apple][s, peripherals and software. This already was unconventional. I wanted assertive salespeople, not shrinking violets, but individuals who could competently explain and represent the product and brand. A theorem in sales says the prospect will tell you as much as s/he thinks you the salesperson understands. The more competent you come across, the more the prospect will open up. For our salespeople the idea was to help retailers sell better and more to their customers, as opposed to the old pressure game of getting our partner to order more, colloquially known as "stuffing the stockroom in the basement". In an old view of the retail trade, getting a dealer to order more was a way to hurt competitors. With the tight finances typical of the retail, if an outlet committed substantial cash to a larger order from Supplier A, this would prevent Supplier B from getting more business.

We played the game differently (I'll avoid the later Think Different phrase here). We combined uniform pricing, speedy logistics and tight financial controls.

We chose an unorthodox practice: no quantity discounts. Our goal was to prevent larger retailers from using their purchasing power to pay less for Apple products and start price wars, which would prevent smaller retailers from delivering full technical services to their clientèle. The Big Dogs (such as retail chain Fnac) didn't like it and, for a while, refused to carry our products. Others scoffed and threatened to order one Apple a day (bad pun, I know), to which we said we'd be happy to be on their doorsteps every morning— as long as they were paid up for previous purchases. That's where managing the financial relationships came in, right behind Michel Delong's speedy logistics. Gilles Mouchonnet's team assigned a carefully considered credit limit to each one of our retail partners. Using an example, retailer S could be 'current' (meaning well within the 30-day payment terms for previous purchases), but we would still decline to ship more merchandise if that would exceed their credit limit. If you wanted more product, you had to pay for what you had already received. In France this wasn't a normal way of doing business. But, after an adaptation period, it worked. I must add that it greatly helped to have a strong brand and desirable products which wouldn't linger on shelves.

We recommended having no more than a week's worth of inventory; pay us quickly and we'd ship the same way. This helped smaller retailers make enough money to supply good service to local customers without fear of being undersold by big chains. We didn't want price wars; we had seen the deleterious effects of the race to the bottom too often in consumer electronics. In practical terms, this was effectively a form of price control that didn't violate French laws.

All this took place in a climate of benign neglect by Apple's European and US management. This let us run slightly unconventional experiments. One was distributing products outside Apple's standard product line. Going back to my encounter with the founders of *Le Chat Mauve*, we financed a first batch of the Peritel interface cards; after a satisfactory trial we put them in our warehouse and supplied French Apple dealers, who uniformly loved their superior color performance. We also worked with the American College in Paris to use their computer facilities to provide a dedicated network, a miniature Compuserve which Apple customers used to exchange mail, news, and information. Naturally we called it Calvados[25]. (There was no Internet then.) Not everything worked — we had trouble, for instance, with an interface to portable typewriters. It didn't sell because it didn't work and made a few people unhappy. My mistake.

Besides the happily effective sales/finance/logistics machine, in the French market, we managed to own the public discourse on our products and, I dare say, on personal computing.

I recounted earlier how, during my first trip to Cupertino, I saw how Apple published an Apple Magazine including a Ray Bradbury poem titled "Ode To The Quick Computer", and another piece from Johnathan Livingston Seagull author Richard Bach. (At the time, Apple didn't explicitly portray itself as standing at the crossroads of Liberal Arts and Technology — Jobs created this framing years later when he came back.) I was thrilled; it spurred me to think of ways to position Apple back in France, an environment that loudly valued cultural pursuits.

Still in Cupertino, sitting at the back of an otherwise uninteresting meeting, I caught parts of what became our standard explanation of what personal computers were for. As we came to repeat on every occasion, these were tools to do five things: Think, Organize, Communicate, Learn and Play. It sounds trite as I write this in 2022, but in the early eighties PC makers discussed important bits and bytes details and stayed there. For reasons I still can't fathom, our competitors let us 'own' the definition statement.

---

[25] https://en.wikipedia.org/wiki/Calvados

Back in France we met <u>Jacques Séguéla</u>[26] and his merry band of thinkers and writers. On our behalf they performed *qualitative* market research for what PC meant in the minds and, more importantly, guts of normal people.

The outcome of such research was couched in seemingly esoteric language. Séguéla's team told us the PC was endowed with three important attributes: Solipsistic, Heuristic and Promethean.

Let's unfold these adjectives to see how they resonate.

Solipsistic refers to a one-on-one, immediate relationship, without mediation, with the computer, as opposed to dealing with an organization. As a gifted but troubled programmer once told me, he liked his computer because it didn't judge him, it obeyed him (most of the time) and, by allowing him to perform difficult, valuable tasks, it gave him a much-needed sense of self-worth. My machine, without intercessors, teachers, judges, or intruders.

Next: Heuristic. As the etymology indicates, heuristic (*eureka*) refers to "good finding". My computer is a rewarding instrument of exploration, of valuable discoveries.

The Promethean attribute deserves more elaboration. It refers to our deeply entrenched need to feel we're masters of our inventions.

Once upon a time we invented symbols such as the letters of the alphabet and Arabic numerals[27]. These symbols gave us boundless power. With symbols we could write Elizabethan poetry, describe Wall Street greed, and produce equations explaining general relativity.

---

[26] https://fr.wikipedia.org/wiki/Jacques_S%C3%A9gu%C3%A9la
[27] https://en.wikipedia.org/wiki/Arabic_numerals

But *our minds* were limited. Our central nervous system left us behind; it stopped evolving while we raced ahead with ever more complex uses of symbols. We had trouble remembering long symbol strings — storing the content of sacred texts in a single brain was an extraordinary skill of exceptional individuals. And we had difficulties performing even simple arithmetic operations in our heads and extracting cube roots. The invention of writing and, later, of the printing press went towards supplementing and expanding the power of our brains. But while that was good for storing, reproducing, and sharing symbol strings; it still fell short when combining, manipulating strings, and calculating. The infinitely flexible symbol manipulation machine arose when the modern computer was born in the forties[28].

Turn the clock three decades forward, and the computer became *personal:* a machine you could lift with your hands, your credit card, and your mind.

We had no need for the people in white coats, the techno-priests who attended the big mainframes. Personal computers were indefatigably patient idiots with the potential for infinite storage and symbol manipulation.

Personal computers made us whole; they supplemented our central nervous system and allowed us to catch up with our invention: symbols. Put in a more poetic and mythological way, they returned to us what the gods had stolen from us. Hence the Promethean attribute.

This was summarized in that easily digestible and repeatable taxonomy of personal computer roles: Think, Organize, Communicate, Learn and Play.

---

[28] https://en.wikipedia.org/wiki/Computer

To this day, I wonder how we came to own this gospel of personal computing, but we did. In mythological and biological terms, we explained the deep *raison d'être* of PCs. And in practical terms, we described the five categories of activities they enabled or enhanced. All this while competitors spent most of their airtime discussing bits and bytes, fun topics for geeks but neither interesting nor meaningful to normal humans, who were more concerned by *what* the PC did for them.

(This is partly unfair. In the best-case geeks who speak the esoteric language of bits and bytes are indispensable early adopters; they are the "bleeding edge", they show the way to The Rest of Us. Luckily, we spoke from both sides of our Apple France mouth, catering to the interests of technically inclined trailblazers, while also doing our best to make sense for people more interested in simply using VisiCalc, AppleWriter or PFS:Graph.)

Again, more by instinct than careful exploration of strategy, we felt we could productively differentiate ourselves using better communication with our retail partners. In 1982, once we had put in place our management team and facilities, we decided to write to our 'constituents' every week. The *Apple Hebdo* newsletter was born. Every Apple France department was to contribute news and updates on their own bailiwick: tech support, repair procedures, sales promotions, tips, and tricks. I would contribute a short introductory message, soon supplemented by a Rick Erickson cartoon. Rick would look over my shoulder, literally sometimes, and draw as I wrote my piece in longhand, to be quickly typed by my assistant, slapped on top of the other documents, and then passed to our printer and routing ally Lucien Ruhier. Lucien was a colleague and friend from my Data General days; he had been the type of Chief Financial Officer who suddenly made me feel smart after I removed his predecessor, whose convoluted 'explanations' made me feel stupid. After Data General days Lucien became the CEO of a printing and routing company which belonged to a cheese conglomerate, the Bongrain group. Said conglomerate wanted to make its own cheese labels and, while at it, sell printing and routing services to others. Lucien and I kept in touch. One day over lunch, as I was describing our expanding activities, he offered to print and route *Apple Hebdo*.

Our newsletter was well-liked, it was inexplicably unique, and it effectively supported our sales strategy of helping dealers *sell* more. They appreciated the support, and perhaps even more that we weren't leaning on them to *buy* more. The beginning was not without hiccups — not every member of the team liked setting time aside to tell Apple dealers what to do to better advance our joint goals. I recall grumblings such as "I didn't join to write weekly missives," but reactions from recipients and competition between departments helped. (Jumping ahead, I was quickly rebuffed by Cupertino marketeers when, after landing there in an engineering leadership position, I suggested Apple US send its own weekly missive to dealers. "You're from engineering, mind your own bits and bytes business." That was after questioning why an ostensibly marketing person like me would run an engineering organization.)

Almost Illicit Fun

Clear positioning, a strong team and a PR genius helped Apple France become Apple's largest business outside the US.

Previously I recounted how we built a service company for Apple retailers. This chapter explains how we reached actual customers by lobbing a simple, resonant message over the heads of those retailers. Here advertising didn't work as the primary medium — the few ads produced by our advertising partners weren't bad, but they failed to communicate the superior life-enhancing identity we were offering.

One such partner was Jacques Séguéla, the advertising magician mentioned earlier, already well known for launching Citroën cars from the deck of an aircraft carrier and contributing to François Mitterrand's victorious presidential campaign. Séguéla graciously repaid our tepid reaction to his Apple ads with a signal favor: he put us in touch with PR guru Lionel Chouchan, the creator of several movie festivals, known for his impeccable integrity (even more remarkable in this business), creative mind and densely rhizomatic[29] address book. Lionel's most notable creation for us was *La Fondation Apple pour le Cinéma* (The Apple Foundation for the Cinema). When he outlined the project, I told him we couldn't afford the cost of such an operation. He knew of a need we could inexpensively fulfill helping young and always cash-strapped movie directors promote their movies. For 30,000 francs (about $20K today), the Foundation could pay for posters on downtown columns

---

[29] https://en.wikipedia.org/wiki/Rhizome_(philosophy)

advertising movies and theater shows. He'd assemble an unpaid jury of actors and scriptwriter friends, take everyone to Martinique (expenses paid) for a few days of relaxation and movie reviewing, and then naming a winner. Even with the prize the costs were reasonable, because Lionel leveraged old festival connections with airlines and hotels.

Sitting once in front of Catherine Deneuve, I felt like a peasant invited to a chic soirée, but it worked. Jim Jarmusch won for *Stranger Than Paradise,* and the publicity we got was worth far more than the operation's total cost of approximately $50K. Even when the first reactions weren't positive, the publicity ended up working for the positioning we sought. At the Cannes Film Festival, I recall a French radio reporter putting a microphone in my face and demanding to know why a computer company was concerned with the movie business. This was a golden opportunity to explain how Apple was different and made tools for creative pursuits rather than instruments of oppression via mechanization of office tasks. The gent chuckled — everyone understood what and whom I meant.

Those were challenging times for Apple. Since 1981 the IBM PC and its cohort of clones had been beating up Apple in corporate America, in office productivity applications with fast machines equipped with hard disks and application software such as the popular Lotus 1–2–3. While in the US Apple marketers were struggling to promote the company as a worthy opponent to Big Blue and its clones, at Apple France we decided a frontal assault would leave us straight to perdition.

Early in our existence we had positioned Apple as providing tools for creative pursuits, only occasionally getting our machines in the forbidden office fortress through the side door, as memorably depicted by Rick Erickson's poster:

GRANDS COMPTES

DEUX APPROCHES

LA LEUR   LA NÔTRE

Image #2: Big Business. Two approaches: Theirs, Ours.

Courtesy Rick Erickson

A word of explanation (Image #2): The top labels say Big Business and Two Approaches.

Theirs: to come as hopeless IBM competitors; Ours: using charm to get the key for the side door.

This worked in a French culture that didn't like IBM, and Apple's California Chic, fueled by Steve Jobs' already mythical charisma, made us the People's Liberation Army from Computing Oppression.

Lionel Chouchan had more ideas. For example, we didn't run a conventional press conference for the Mac launch. Instead, drawing again on his media contacts, we put a mock TV evening on a Paris theater stage, complete with celebrities' videos, demos, yours being roasted, and musical numbers by Gérard Lenorman and Michel Sardou, well-known singers in France. Together, the medium and the message worked well; to the French mediasphere, they successfully reinforced our creative maverick identity.

The excitement got to us, in a good way.

One day I overheard a loud telephone argument between a service tech and one of our customers. A logic board had been damaged because of an interface card's forceful (as in mallet traces on the edge) backward insertion. The customer didn't like being told he or, more likely, his offspring had abused the hardware. The pin, whose function was to prevent such bad insertion, was deemed too weak, and legal action was threatened. The next step was easy to see. I inserted myself into the heated exchange and asked the unhappy customer if he wanted us to buy his system back because "we couldn't afford a single unhappy Apple patron." The answer was a resounding **no** — he wanted to keep his Apple][. Next, I arranged for the customer to bring the dead system to our service department. As a special favor, we'd recycle a board from an otherwise dead machine and bring his machine back to life. Lastly, I asked for his child's age. Why? For the t-shirt, of course (Apple t-shirts were hot items then); I wanted to thank him for the opportunity to make things right.

From then on, we made giving t-shirts to complainants a standard but unofficial practice.

Behaviorists might argue we were naively encouraging questionable behavior and exposing ourselves to a deluge of ill-founded complaints. No. There are only very few really dishonest customers, and one can often see them coming. It's better to have a reputation for being nice, if occasionally naive, than as a martinet. And the positive word of mouth was worth much more than the t-shirts or cost of the deniable repairs.

One last thing.

As business grew nicely, phone traffic became a problem, with callers complaining about long waits — such as five to ten minutes. As it happened, we had asked a radio personality, Kriss Graffiti, to record the translation of audio cassettes supplied with the original Mac, intended to familiarize customers with the strange new machine. She went over the budget she had quoted, expecting to have to eat the extra recording studio expense incurred. She was pleasantly surprised when we didn't argue and settled without ado, despite her expectations after years in the business.

A pleasant conversation ensued where I mentioned our telephone wait time problem. Because she had a seductive, clear radio voice (I use the past tense because tobacco-related cancer killed her), my dream was to have her read stories to callers as they waited. She immediately accepted, told me she'd do it for free as a thank you, and said she'd write the stories herself. These were sweet little tales that markedly changed the mood of callers. Some even asked to be put back on hold because they wanted to hear the end of the tale.

These were fun times, and our numbers contributed to the mood. We became Apple's largest business outside the

US.US execs liked to end European tours in France because we lifted their spirits.

Unfortunately, the much-awaited and much-delayed Macintosh quickly got in trouble, especially in the US.

Mac Hopes and Troubles

The Macintosh had a difficult birth: late, expensive, poorly positioned. Its problems were to become my opportunity.

Rewinding the clock…

It's November 1983; I'm sitting in the auditorium at Apple's worldwide sales meeting in Honolulu. The house lights dim, and "1984" begins (Image #3). Conceived by ad agency Chiat/Day, directed by Ridley Scott of *Blade Runner* fame, and destined to be aired nationally only once (during the 1984 Super Bowl), the minute-long movie leaves us with this message:

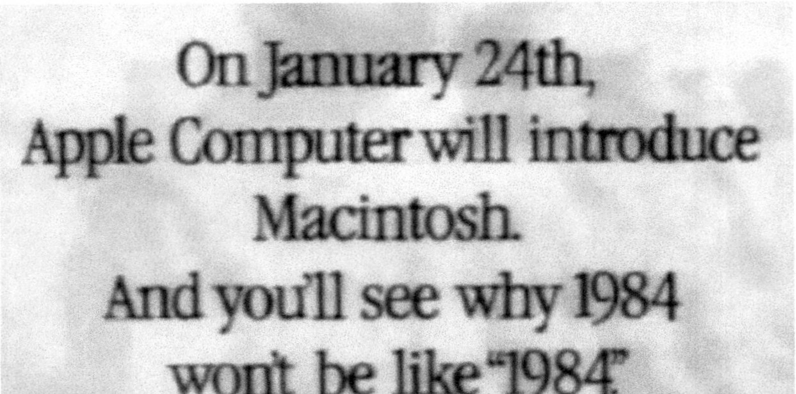

On January 24th,
Apple Computer will introduce
Macintosh.
And you'll see why 1984
won't be like "1984"

Image #3

The lights come halfway up. Steve Jobs' magical brainchild is lowered from the flies, *deus ex Macintosh.* Halfway through its descent the Mac boots up, and we hear the newborn's wail, the now familiar Bong.

To this day I remember the electrifying effect on the audience, and troubled thoughts regarding mass persuasion — criticized on screen and then, without irony, performed right in that room.

Apple's assembled sales organization was delighted by the Mac's enchanting presentation, its (almost) never-before-seen user interface. But there was a nervous energy under the surface: Would the Macintosh save Apple from the IBM PC and its clones?

Born in 1981, almost three years before the repeatedly delayed Mac, the 16-bit IBM PC had made mincemeat of the 8-bit Apple][(and the troubled Apple ///). With the introduction of the PC XT and advent of Lotus 1–2–3 in 1983, the competitive situation had become even more severe. The XT sported an integrated hard disk, something Apple machines lacked. Lotus 1–2–3, the model for integrated software, was rightly called a "killer app" because it displaced products such as VisiCalc and, by extension, machines such as the Apple][that didn't run the new champion.

Image #4

From such a stodgy company, IBM's PC marketing was shockingly clever. They appropriated Charlie Chaplin as a mascot and ran a successful TV ad campaign that positioned their machine as the personal computer. I was stunned: They're stealing our song! (Image #4)

Apple replied with cheeky "Welcome, IBM" ads (Image #5):

Welcome, IBM.
Seriously.

Welcome to the most exciting and important marketplace since the computer revolution began 35 years ago.

And congratulations on your first personal computer.

Putting real computer power in the hands of the individual is already improving the way people work, think, learn, communicate and spend their leisure hours.

Computer literacy is fast becoming as fundamental a skill as reading or writing.

When we invented the first personal computer system, we estimated that over 140,000,000 people worldwide could justify the purchase of one, if only they understood its benefits.

Next year alone, we project that well over 1,000,000 will come to that understanding. Over the next decade, the growth of the personal computer will continue in logarithmic leaps.

We look forward to responsible competition in the massive effort to distribute this American technology to the world. And we appreciate the magnitude of your commitment.

Because what we are doing is increasing social capital by enhancing individual productivity.

Welcome to the task. **apple**

Image #5

I felt this was twice mistaken. Not only did it make us sound presumptuous, but why were we spending money advertising the opposition's ware? (I later realized that I was wrong on the former point: It was a good idea to position Apple as an equal — and precursor — to IBM.)

There was also Job's more defiant pose, not an ad but widely reproduced (Image #6):

Image #6

Photo by Jean Pigozzi—

As released publicly by Andy Hertzfeld

Good fun, probably intended to motivate Apple troops, but not necessarily endearing to prospects in conservative businesses.

Still, initial reactions to the Mac were positive. Here's a young 'Lawrence' Magid in his hot-off-the-press review:

"I rarely get excited over a new computer. But Apple's Macintosh, officially introduced last Tuesday, has started a fever in Silicon Valley that's hard not to catch. My symptoms started when I talked with some devotees from Apple and the various companies that produce software, hardware, and literature to enhance the new computer. I was hooked when I got my hands on the little computer and its omni-present mouse. Apple has a winner."

While the excitement was genuine, sales of the original 128k Macintosh were hobbled by its relatively high price ($2,495) and the same lack of features that hurt the Apple][: no hard disk, little office productivity software and definitely no Lotus 1–2–3. Although it found favor with a few creative professionals and educators, the Mac was shunned by "Corporate America", which stuck with IBM-compatible machines running Microsoft system software and apps.

By early 1985 Mac sales still weren't taking off, and sinking sales of the Apple][would lead to the shutdown of the company's Texas manufacturing plant and the company's first-ever layoff. Something had to be done.

With a foothold in the "creative space" and education applications, Jobs thought we could sell the French government on having a large local company such as Thomson take a license to build Macs and sell them to the education market, creating a success story and fatter numbers.

Context is needed here. On December 31st, 1981, President François Mitterrand offered his New Year wishes on prime-time TV. They included *L'Informatique Pour Tous* (Computing for Everyone), the brainchild of influential publishing family scion and visionary/author/agitator/brief cabinet minister Jean-Jacques Servan-Schreiber. JJSS, as he was known, founded the weekly *L'Express* in 1953 and authored the strongly worded *"Le Défi Americain"* (The American Challenge), an essay which quickly sold 600,000 copies when it came out in 1967 — an enormous number for a political essay.

I recall how I felt when Mitterrand expressed his vision of Computing For The People: This is the pitch for Apple! We promptly and efficiently took up the refrain. Luckily our US masters were launching a Kids Can't Wait marketing blitz targeting the education market. We piggy-backed on it, called it *L'Avenir N'attend Pas* (The Future Can't Wait), exploited government regulations again, and sold beaucoup Apple][machines and the color monitors we had had "made to measure" by Philips Italy. (The monitors were an idea from Michael Spindler, the recently departed and much-missed friend who was then European Marketing Chief.)

Complicated conversations with politicians, world-renowned parasites, and French industrialists went nowhere. Still, I got to know CEO John Sculley, brought in by the Apple board to provide "adult supervision" to Jobs. Sculley gave our staff a talk describing Apple's future. It was the best business talk I had ever heard, and I told him so. I wasn't flattering him — it was my honest feeling, and my hope for the company's coming years.

Back in the US things were becoming tense. To counter the Mac's perceived and real weakness in business productivity apps, Jobs came up with the concept of a Macintosh Office, including a Local Area Network (LAN), a File Server (essentially a networked hard drive), and a networked laser printer. This was vintage Jobs: A grand vision supported by a spectacular demo. Unfortunately, it was only a demo.

Deploying the kremlinology that got me hired to start Apple France (see Chapter *"Turning Point 2: The Apple Impact"*), I sent a pair of notes to Sculley, in which I dissected Jobs' story. The Mac Office concept would never become a reality (I wrote), and even if the fantasy could be true, it wouldn't solve our corporate America market problem.

My memos were not universally well-received, but neither was the 1985 Mac Office. (However, the LaserWriter and the AppleTalk LAN were later to become key components of Apple's successful desktop publishing push.)

In parallel, Apple France numbers kept improving. As they did, to keep a sober perspective, I kept reciting to myself an old La Fontaine fable: *The Donkey Carrying the Image*[30] (In French: *L'Âne portant des reliques*[31]) In the fable, as the procession went on, the poor donkey thought the incense thrown at a religious relic was for him. In my case, I couldn't afford to forget I was carrying the Apple image on my back. That said, our numbers validated my predictions: We soon became the company's biggest revenue and profit generator outside the US. As indicated in an earlier chapter, because of this US management envoys habitually ended their European tour in France to shore up morale after visiting less happy places. The Mac had an auspicious beginning in France.

---

[30]https://portalbuzzuserfiles.s3.amazonaws.com/ou-19371/userfiles/files/vls%20aesop%20fables/the-donkey-carrying-the-image.pdf

[31] https://fr.wikipedia.org/wiki/L%27%C3%82ne_portant_des_reliques

Macintosh's Early Days in France

French politics aimed at taking control of the Mac helped Apple.

We had a grand time launching the Mac in France. It coincided with efforts to escape the Seedrin pretense. Everyone knew Apple had distribution, retailer, and software developer support in France, but as I explained earlier, it wasn't an Apple-owned company because French authorities wouldn't allow Apple to capitalize a business in France. In their minds keeping Apple money out of France protected local PC makers. Our good health made this more ridiculous every month, and a decision was made to allow an Apple investment if the Cupertino company would launch a technical activity benefiting French industry. We offered to start a software development office, agreed on details, and now needed to find a proper location. I discovered that 'proper' meant offering a political advantage to a friend of the current administration.

I was directed to Jean-Marie Rausch, mayor of Metz, member of France's Senate and region president. Sitting next to him at lunch, I listened to him mention his troubles with his PC's 'keyboard driver'. A geeky (middle-aged) politician! I offered to bring him a Mac the following week. On the agreed day I put a Mac in the trunk of my car and drove the 200 miles from Paris. When I tried to alight at the mayoral building, a cop wanted to wave me off. I politely explained I wanted to drop something off for the mayor. When I took the Mac out of my trunk, the cop knowingly exclaimed: "Ah, a Mac for the mayor, he'll like it!" I was in Apple-friendly territory. An agreement was blessed by all parties, and Apple was now official in France. We easily met our commitment: the phrase "Software Development" covers anything that involves writing lines of code, which could include adapting US software.

Then disaster struck — in the US. Mac sales failed to reach expectations, causing Apple — in this case John Sculley and Steve Jobs — to look for ways to stimulate sales and generate favorable publicity. As recounted earlier, after much agitated brainstorming, the idea came up to make a deal with the French government to start a factory in the country to manufacture Macs under license and funnel some of its production to France's schools.

Here, I need to back up and explain the context for this 'creative' (crazy!) suggestion.

In yet another stroke of luck, French President François Mitterrand, elected in May 1981, advanced *Informatique Pour Tous* (Computing for Everyone) in his New Year's Eve address on French TV. Mitterrand was singing our song, expounding the need for kids to learn computers. As one easily imagines, this 65-year-old very traditional politician had no intellectual or emotional connection with computers. The *Informatique Pour Tous* idea came from Jean-Jacques Servan-Schreiber, more commonly known as JJSS, politician, member of a publishing dynasty, founder of weekly *L'Express,* and author of best-seller *The American Challenge*. JJSS saw the future of computers for everyone, and the opportunity for Mitterrand and his administration to ride the coming wave (described in *Chapter "Mac Hopes and Troubles"*). Mitterrand agreed, and his government funded the creation of the *Centre Mondial Pour l'Informatique et La Ressource Humaine* (The World Center for Computing and the Human Resource— note the singular.) The Centre Mondial leased a beautiful building on Avenue Matignon, the elegant former Time Life building, and imported US academic celebrities such as Nicolas Negroponte and Marvin Minsky. Because of JJSS' often flamboyant pronouncements, the Centre Mondial met with occasional derision and criticism for its budget and importation of US researchers, but it nonetheless did a terrific job promoting personal computers in schools and homes. A few French personal computer — or, as they said then, microcomputer — makers such as R2E's Micral, Thomson and its TO7 or SMT's Goupil got a small boost, but the main beneficiary was Apple. We were more visible; we had a coherent and culturally compatible Franco-Californian story, and a well-supported distribution network. At about the same time Apple US launched an Apple][promotion titled Kids Can't Wait to offer educational discounts. We ran a local adaptation called *L'Avenir N'Attend Pas* (The Future Can't Wait) and offered a 50% discount to education and vocational training non-profit organizations. French law was such that anyone could start a so-called Loi (Law) 1901 educational association; a single person could file such a declaration. Crafty French customers jumped on the opportunity, and we soon enjoyed a marked increase in Apple][sales to a motley crew of education philanthropists. Apple Europe management had mixed feelings about this 'creative' program, one that, in conservative eyes, stretched the spirit of the US program a little too far. But our noble and worthy overseers liked the numbers and the good PR, especially compared to other struggling parts of the company.

Back to flagging Mac sales in the US: JJSS contacted Jobs and suggested a licensing agreement leading to a Mac manufacturing plant in France. The output would be mainly, but not solely, directed to France schools[32]. Informal negotiations started via international lawyer Samuel Pisar, another member of Mitterrand's coterie. For a while talks heated up. Steve Jobs and John Sculley came to Paris; Steve met Mitterrand at the Élysée palace. In preparation for follow-up negotiations, we set up a war room at the nearby Hôtel Bristol. There France Télécom did a double take when seeing a Mac — inexplicably those tech execs had never seen one. To them, the top and back of Steve's creation was a too-close-for-comfort copy of one of the Minitel boxes they sold to French consumers:

(I didn't tell them that, three years earlier, Steve had asked me to ship a Minitel box to Cupertino. The resemblance was a true coincidence as, by 1981, the Mac box shape had long been 'frozen', I had seen one at Microsoft in Redmond by then. The coincidence must have arisen from the need for a handle to carry the similarly sized enclosure.)

Another member of Mitterrand's circle, Gaston Defferre, mayor of Marseille, got involved and offered Jobs an overflight of the city to show possible sites for the Mac plant. But negotiations went nowhere. Big tech industry players saw such a plant, "French-owned" or not, as setting the fox in the chicken coop. Also, a plant would need a couple years to be permitted and built, and the manufacturing machinery would then need to be installed and debugged. Meanwhile the French government would need a substantial shipment of machines from Apple's Fremont plant in California to install Macs in schools. There seemed to be no money in the country's budget to pay for tens of thousands of machines. "No problem," said Samuel Pisar when I tiptoed around the

---

[32] https://en.wikipedia.org/wiki/Apple_II_clones

payment question, "We'll do a barter deal for Bordeaux wine." That told me we had been in fantasyland for a while.

Negotiations failed but the leaks did wonders for our PR, much to the indignation of local competitors. Once again, and not for the last time as Apple historians know, the Steve Jobs magic had worked, making Apple look much bigger than it actually was, especially when business wasn't doing well — both Apple][and Mac sales were failing to meet expectations.

I was getting restless despite the successes of Apple France. Many business executives would have been thrilled to keep running such an operation with a happy and competent management team. Not me. Nothing rational. I had an itch: I wanted to go to the US and fulfill an old dream.

Digression: This was a long-standing desire that started during my HP years when, around 1973, at a visit to the Loveland division that made the company's desktop computers, I asked its engineering boss, Bob Watson, if there was any possibility I could come and work there. Nothing came of that conversation, and I left HP in 1974. Even though I exhibited some dexterity with products, I later realized that I lacked the engineering training to join Bishop Bob's team (Watson was a Mormon bishop). I would meet him again in 1989, when running Apple's engineering operations in Cupertino. We discussed and failed to agree on a contract manufacturing arrangement for printers. Bob worked for Dick Hackborn, the maverick head of HP's inkjet printer division in Boise, Idaho. Years later, in 1996, we again met at Be when I hired Dick's daughter Diane as the engineer in charge of BeOS Applications Frameworks. She gained a solid reputation for her skills and went on to perform similar work at Google.

It looked like I might have an opportunity to go and work in Cupertino. The Mac was caught in a vicious circle. It didn't have enough applications, which hurt sales. Low sales meant software developers lacked motivation to write software for Jobs' brainchild. So, I offered to come and build a Software Division. I even wrote what I thought was a form of business plan for it, using Dave Winer's *Think Tank* "idea processor" (outliner). In retrospect I shudder: my draft proposal was a terrible idea, a stream-of-consciousness disgorgement, a Theory of Everything on the topic. I presented it to John Sculley, who mercifully ignored it. The Software Division needed no such *logorrhea*; its *raison d'être* was simple: break the vicious no-application-software/no-hardware-sales circle.

There was a bigger problem, at least from my standpoint: Steve Jobs, a legendary control freak, wanted the Software Division to report to him. Knowing Steve's management style, and aware of my limited control over my own emotions, I didn't want to work for Apple's mercurial visionary. Nothing happened for a while. The following chapter will examine how the stalemate was resolved.

AMERICAN DREAM (SORT OF)

Hard Landing in Cupertino

Steve Jobs was forced out of the company — an event that later proved a stroke of luck for him and the company he co-founded.

As we saw in *Chapter "Mac Hopes and Troubles"*, Steve Jobs did a good thing by welcoming IBM to the personal computer market. Regrettably, it turned out to only be half a good thing.

Let me explain.

Jobs' good idea was to put Apple on an equal footing with IBM, or at least to make the smaller Cupertino company look like David against Goliath.

But David sling held no slayer stones. On the contrary Big Blue's[33] PC featured a color screen, quickly got "killer apps"[34] such as Lotus 1-2-3 and starting with the XT model had a built-in hard disk— none of which the Mac offered. And IBM's PC was soon followed by a proliferating crowd of personal computers running the same software; they were called PC compatibles or clones[35]. In 1982 Bill Gates traveled around with such a portable computer from Compaq.

At the time IBM lorded over corporate America. As the saying went, you'd never get fired for buying IBM. In strait-laced business circles Apple's maverick image, Steve Jobs' statements and the famous 1984 commercial featuring Big Brother as a thinly veiled reference to IBM lording over passive

---

[33] https://en.wikipedia.org/wiki/Big_Blue_(disambiguation)

[34] https://en.wikipedia.org/wiki/Killer_application

[35] https://en.wikipedia.org/wiki/IBM_PC_compatible

crowds of workers were all borderline offensive to the very people Apple wanted to seduce.

As stated earlier, seeing Macintosh sales floundering, Steve hurriedly put together a concept and a demo: Mac Office[36].

In an office, an Apple File Server would hold the files shared by collaborating knowledge workers — his demo graphically explained movement of information through a Local Area Network later, called AppleTalk, which also included a LaserWriter printer. Demonstrated internally in the fall of 1984, Mac Office was publicly announced in early 1985, supported, if that's the right word, by an ad titled Lemmings[37] which satirized people who blindly follow crowds into buying You Know Whose computers.

At the time of the internal demo and after the public announcement, I wrote two white papers addressed to John Sculley denouncing the demo as just that, not a product likely to ship in any foreseeable future.

My Cupertino masters got tired of my relentless kibitzing about the company's Hail Mary shots, particularly my kremlinologist dissection of Mac Office and its File Server. This earned me a ticket to Cupertino and an injunction to leave the peanut gallery and, instead, come a hand.

As indicated earlier, Initially, I was to work on a putative "Software Division" — we didn't say App Store, the Internet wasn't in use outside research labs yet. The idea was to encourage third-party developers to write software for the Mac and help them make a living doing it.

For the business market, the PC (IBM's and the rising clones) had a 'killer app' in a popular program called Lotus 1-2-3 that

---

[36] https://en.wikipedia.org/wiki/Macintosh_Office

[37] https://en.wikipedia.org/wiki/Lemmings_(advertisement)

combined spreadsheet, graphing and database functions. The Mac badly needed a response. It needed software to make it more useful to more people, it wanted more applications that made obvious the Mac's distinguishing graphic UI and power.

I was to report directly to CEO John Sculley, but Steve Jobs, in his trademark posture, would have none of it. He would not loosen his grip on the Mac. Caught between the two, I was adamant: I couldn't work for Apple's mercurial and controlling co-founder. Previous trips to Cupertino, along with Jobs' visits to France, had convinced me I didn't (yet) have the temperament or inner security to 'collaborate' with him. It would take a 1993 brush with death, a damaged carotid artery and a stroke, my wife's love, couples therapy, more than a decade of Zen meditation and a series of helpful failures for me to finally grow from a recovering assoholic to a happier and more helpful state.

In the spring of 1985, with the Mac business in serious trouble, Jobs sensed his future was at stake. He brought things to a head by approaching members of Apple's exec team, asking them to side with him and run Sculley out of town.

I have a vivid recollection of a key moment. We were having pre-dinner drinks at the house of Al Eisenstat, Apple's General Counsel, and John Sculley was to leave for a sort of state visit to China the next morning. Seized by the moment, I put my index finger on John's breastbone and told him that if he left for that trip, he'd lose his job. After I explained why, John made phone calls and stayed. The coup failed, and Jobs was deposed. In the reorganization that followed, I became VP of Product Development.

Jobs' ouster was controversial at the time; for some it still is. In a 2011 email to the *New York Times,* Larry Ellison gratuitously damned the move in a discussion of HP's latest exec shuffle:

"The HP board just made the worst personnel decision [firing then CEO Mark Hurd] since the idiots on the Apple board fired Steve Jobs many years ago".

Ellison was a close friend of Jobs and makes provocative statements, but many critics share that sentiment. The article containing the above quote agrees that the move was "one of the most infamous board blunders in tech industry lore". Even Sculley, in remembrance of Jobs' passing in 2011, said, "in hindsight, it was a terrible mistake".

I disagree. Jobs ultimately succeeded because he was fired. Jobs' amazing turnaround of Apple when he returned in 1997 has led to an understandable but causally backwards revision of history. With due respect to Sculley, hindsight shows that Jobs' dismissal was a stroke of luck of historic proportions: for Apple and shareholders, the industry, Jobs himself, and especially us later Apple customers — users of his often-magical creations.

The Jobs I knew in 1985 had effectively no experience outside Apple. Bill Gates once said, "success is a terrible teacher." (The French translation, *maîtresse,* both master and lover, is even better because it combines knowledge with infatuation.) The success of the Apple][might have seduced Jobs into believing he knew while he might have simply been a kind of Chauncey Gardner: at the right place at the right time. That's excessive. Despite his lack of experience outside Apple, Jobs had an indisputable combination of aesthetic flair and intuition, the ability to smell opportunities, as he demonstrated when he zeroed in on Xerox Parc's Alto user interface.

Most important was Jobs' charisma, the ability to make you believe his story despite evidence to the contrary — the SPJ (Steven Paul Jobs) Reality Distortion Field. And there we see the real problem: self-induction. As with electromagnetic fields, the Reality Distortion Field also worked on the emitter, who came to believe with utmost and dangerous sincerity. Jobs was susceptible to being taken in by his own tales.

Like all heroes in mythology and Hollywood movies, Jobs was wounded. He'd been thrown off his high horse. But like a true lionheart, he brushed himself off, started NeXT and bought an almost-abandoned technology team from Lucasfilm to become Pixar.

What did Jobs know about the complex, treacherous movie industry or animated films? Nothing. Yet he turned that team into an astounding success, starting with the 1995's unforgettable *Toy Story*. Pixar was later sold to Disney for $7.4B. If all he did in life was Pixar, Jobs would still be rightfully remembered as a titan of industry.

NeXT was a different story. It was a technical success but not a commercial one. When Apple came calling in late 1996, looking for a better option than acquiring BeOS to replace the then-ailing Mac OS, Jobs wasn't interested. NeXTStep, the company's foundational software, had lain fallow for some time; the company was focused on WebObjects, a set of software tools aimed at helping businesses develop e-commerce and other web-based applications.

But, according to one insider, others in the company saw opportunity. NeXTStep was quickly brushed up, a convincing SPJ demo was scripted, and Apple bought it for the low price of $429M — a strong candidate for the best bargain in computer history.

Jobs quickly got Apple's board to put him back at the helm, ousting CEO Gil Amelio, to whom we owe lasting — if sardonic — gratitude for getting Jobs back into the company. What happened next was a series of bestsellers — particularly the iPod, iPhone, and iPad. It was an astonishing turnaround, not just for technology history, but for any industry.

Being fired proved a terrific, not terrible, *maîtresse*.

Back in 1985 I found myself in a dangerous and paradoxical position. I, too, lacked experience. I had never run an engineering organization; my knowledge of computer technology was largely acquired on the job, and on weekends of torturing hardware and software and reading manuals. Yet, as VP of Product Development, I was handed the reins to Apple's engineering organization. I benefited from the fact that no one on the exec team had computer engineering experience — many came instead from consumer goods companies such as Pepsi, J. Walter Thompson, and Playtex.

Even worse, I inherited two large organizations which hated one another.

The Mac group was ailing but thought of itself as far superior to the 'traditional' (stodgy) Apple][engineers, calling them bozos and other charming names. The Apple][people thought the Mac folks were a bunch of arrogant, ungrateful bastards. After all (they said), the Apple][was paying everyone's salaries; the Mac was still a pretty demo.

My mission was simple — at least to state: Get the Mac out of the ditch and create a cohesive organization to unite the engineers.

As we'll see, culture rather than technology, and my own emotions, were my greatest challenges.

Cupertino Culture Trouble

Moving from Apple France to Cupertino, and from running a distribution company to heading Apple's engineering, proved more challenging than I naively expected.

I land in Cupertino in May 1985 with three strikes against me:

I'm a true Parisian.

I've been the head of a successful team running Apple France.

I have no experience running an engineering organization.

Parisians don't mince. We don't say, "I love what you're doing, but maybe we can tweak it," or "This is interesting, can you tell me more?" Instead, an engineer calls a colleague's work a pile of canine dejection. The colleague responds by questioning the critic's mental or physical organs. No physical violence, this is merely the opening ritual. After a few more volleys of decreasing intensity — and once their respective credentials have been asserted — collaboration ensues with no bitter feelings.

At Apple France, I could question anything about our work, all the way up and down the entire organization, even if that meant that I would occasionally be sent back to my corner by a loyal, competent team as vociferous in their retorts as I in my critique.

You see what's coming. In Cupertino I fall into the classical trap of continuing to do one's past job in the new one.

My first interactions with engineers — discussing their work, observing a demo — aren't very productive. It isn't just that my unfiltered comments aren't well received, but that the lack of engagement disorients me; no ritual give-and-take ultimately clears the air. If I 'offer' a Parisian-style critique (the French don't give 'feedback' they perform live dissection) dead silence ensues. It's as if an invisible steel curtain has descended between the engineers and me.

The ever-present HR parson (Freudian typo gratefully accepted) attached to the new alien suggests I ask questions rather than give immediate and unmediated feedback. But I object: these people worked with Steve Jobs, so surely, they've been subjected to torrents of dismissive feedback. Yes, said the beard-stroking cleric, but you're not Steve Jobs. They wonder why a marketing person now leads Apple's engineers.

A marketing person, does he mean me? Ah, the convenience of specious taxonomy. Apple France was a sort of marketing organization, ergo I'm a marketing person.

I change my tune.

First, I offer credentials and my new role. I'm a college dropout with a junior Math and Physics degree; my computer tech skills were picked up out of interest rather than necessity over 17 years at HP, DG, Exxon Information Systems and Apple itself. I will not be a boss who tries to out-engineer the engineers — their superior tech chops are a given. Instead, I encourage them to enlighten me.

Following the HR rep's suggestion, I decide to ask questions. Actually, a single two-part question: What are you doing, and Why?

For the What, I rely on a variant of my old I Can't Be Stupid gambit: If I don't understand what you're saying, either you don't know what you're talking about, or you're withholding something. As for the Why, please don't say you're following orders — I know you have a mind of your own, so don't hide behind "marketing demanded it" as if you respect them. Tell me how your work serves our common purpose. Does it improve performance, reduce cost, increase reliability, or forge a killer feature? If you can't tell me, perhaps you should go back to your desk, gather your thoughts, and come up with some answers that make me feel smarter and safer.

As news of these interrogations spreads the ever-present HR kommissar gives me a tepidly reproachful lecture, but it works. The engineers and I quickly reach an understanding of our respective roles and how to productively discuss our work. Perfect? No. But after recalibration, my interactions with engineers are the best part of my five years in Cupertino. I know I'm accepted when off-color jokes are prefaced with: As JLG would say…

The not-best part combines past habits of the heart and mind from my Apple France role, my own insecure, asshole behavior and what I perceive as the fearful incompetence of the marketing side of the house.

I have repeated arguments with marketeers who came to Apple from consumer goods companies such as Playtex and Pepsi. They exude a "We're doomed" attitude, assuming personal computers will soon become commodities. It's an endgame pessimism: cash your bonus, update your resume, and move on. My view is that we're beginning, there's monetizable differentiation to be had and we mustn't race to the bottom.

I also vociferously disagree with marketing's obsession with 'winning' corporate America through a frontal attack on IBM and its clones. We are doomed if we take that approach — the opposition is too firmly entrenched. I argue that Desktop Publishing[38] is a perfect example of a "side door" approach, ready-made for Apple and the Mac. We can woo the creative

---

[38] https://en.wikipedia.org/wiki/Desktop_publishing

types without having to storm the fortress, a metaphor brilliantly illustrated by Rick Erickson during my Apple France days, as presented in the chapter *"Almost Illicit Fun"*.

As for desktop publishing, it was an extremely valuable Steve Jobs legacy. With Adobe[39] and Canon he created the LaserWriter[40], using the PostScript printing language to produce stunning high-quality documents. Aldus, later acquired by Adobe, created an application with a self-explanatory name, PageMaker[41], which allowed 'normal' Mac users to create stunning documents — hitherto the exclusive province of professional graphics shops.

My side-door strategy suggestion falls on deaf ears. Not only are my past ventures in field sales dismissed as irrelevant because they took place in an alien country, but I'm now seen by the marketing folks as a mere techie who wandered out of the labs. I can't win.

Undeterred, and unhelpfully, I couch my objections in less-than-diplomatic terms. I accuse my marketing colleagues of thoughtlessly trying to shoehorn their expensive business school enlightenment, the then-famous Case Method[42] into an incompatible reality. (Unfortunately, 'accuse' is the correct word. I'm not flogging myself here, especially contemplating past actions in the gentler light of today's laboriously acquired peace of mind.)

My wide-ranging disrespect earns me a few enemies, some smarter than others. One particularly perceptive individual tries to set me straight when I protest that I respect his role. "Indeed,

---

[39] https://en.wikipedia.org/wiki/Adobe_Inc.

[40] https://en.wikipedia.org/wiki/LaserWriter

[41] https://en.wikipedia.org/wiki/Adobe_PageMaker

[42] https://www.hbs.edu/mba/academic-experience/Pages/the-hbs-case-method.aspx

you respect the function, but you show too much disdain for the person."

I wasn't ready for the lesson. Because of too many unreturned phone calls, of telling the founder of a dominant software company what to do to himself if he fails to renew a software license for the Apple][and for other needlessly dismissive actions I am temporarily demoted. A few more offenses and I'm almost sent back whence I came.

But the engineers' strong work saves me.

In March 1987 we announce the Mac II; the Open Mac promised on my license plate; and the Mac SE, an evolution of the original Mac with an internal disk drive. (That machine earns justified gibes for its noisy fan but sells well nonetheless.)

The Mac II is well received — it definitely bumps the Mac out of the ditch and presages a series of similar machines. Some, such as the Mac IIci, are a bit smaller; others like the Mac IIfx are considerably more powerful.

As a result, my excesses are forgiven — and sometimes boasted about once I'm back in (most) everyone's good graces. I gain a fancier title and slightly broader responsibilities, including a Research organization.

The last two years had been both productive and emotionally difficult, as I sometimes wondered if I would accomplish my stated mission. But working with Apple's engineers gave me some of the most rewarding moments of my professional life.

Getting the Mac out of the Ditch

I had a few simple ideas for short and medium improvements to help the Mac and was happily surprised by the support most of these ideas received inside the engineering community.

I count 30 faces around the conference table. That's one too many. As Apple's newly appointed VP of Product Development, according to a temporary org chart, I start with 29 direct reports. Who's the interloper?

"Ah, I'm your Human Resources representative. Given the potentially tendentious nature of this meeting, HR felt it would be prudent to attend as an aid to communication".

The meeting did indeed look to be contentious. I needed to harmonize the Mac and Apple][engineering groups into a single euphonious choir. The two factions are openly disdainful of each other, so inevitably some attendees will not be thrilled with the new org chart.

But HR's presence was helpful, although not as intended. The barely hidden smirks and eyerolls (on both sides of the table) were reassuring as I looked around the room. However much these engineers despised each other and feared for their positions, we shared a common target. Engineers liked to call HR the Thought Police, KGB, and Gestapo; they insisted these were meliorative nicknames because those are all effective organizations. To be fair, Apple HR would prove a boon when helping my family and me get settled. Unfortunately, they also created political turmoil as they overplayed their power-behind-the-throne role at the highest levels of the company. Even more regrettably, I allowed myself to get sucked into the turbulent politics that preceded and followed Steve Jobs' departure.

Having formed a (somewhat) unified engineering team, it's time to get down to business.

It's May 1985: Apple][sales are falling and the Mac has yet to take off. We need to make changes quickly to attract new customers and keep the old ones coming back.

First I felt that the Apple][family was becoming needlessly complicated. As a multimedia machine, the proposed Apple IIGS ('**G**' for graphics and '**S**' for sound) was undeniably superior to earlier Apple][s and the current Mac, but it used a virtually unknown processor with a doubtful future — a portent I later failed to heed when starting Be. In addition, the new and unproved CPU chip offered so-so compatibility with earlier Apple][models. So much that Bill Gates, after a demo and presentation, told us Microsoft wouldn't adapt old Apple][software or write new applications for the IIGS. I wondered, a little too loudly: Why are we going down this path?

I paid a political price for my desire to deemphasize the product. The Apple IIGS was the favorite child of a senior exec who took my view as a personal attack (admittedly, I could have expressed my opinion a bit more diplomatically). Bill Gates's rejection of the IIGS should have helped my position, but the ego damage was done and would later almost get me fired when my all-around combativeness made several senior execs wonder if they could or should work with me.

For the Mac I suggested we needed to do three things:

- Implement a few quick improvements to make the Mac feel more muscular,

- Design an 'open' Mac with interface card slots and color display capabilities,

- Slide a robust operating system kernel under the existing Mac OS.

The "muscular improvements" made their way into the next Mac offering, the Macintosh Plus, announced in January 1986. The most important new feature was the incorporation of a SCSI[43] connector, which let customers plug in external hard drives. Less conspicuous but substantial were double-sided floppies, a larger ROM, and increasing internal memory (RAM) to 1MB — an excess which Bill Gates dissed, arguing that nobody needed more than the DOS PC's 640K upper limit. The positive reaction to the Mac Plus was a welcome stopgap. Inside Apple we liked to think that Gates' reproof made the machine more popular.

I expected stiff resistance to 'opening' the Mac but found none. But I offered an argument against my own position: "I realize that Steve Jobs strenuously opposed such features". "He sure did, but that was him and then."

Image #7

---

[43] https://en.wikipedia.org/wiki/SCSI

As a *"cri de guerre"*, I got a suitable license plate for my car (Image #7) and handed out the obligatory t-shirts (Image #8):

Image #8

The team quickly settled on a bus standard and went off to create the Macintosh II. Partly based on the design of the ill-fated Macintosh Office File Server, the Mac II shipped less than two years later.

Unfortunately, my suggestion that we introduce a kernel into the Mac OS was unsuccessful. Here, a brief definition may be helpful — with my apologies to genuine computer scientists for the oversimplifications.

Today, all general-purpose computer operating systems are regulated by a kernel, a base layer of software which sits right above the processor or processors. The kernel combines traffic cop and nanny, ensuring that applications play nice with one another, that hardware resources such as memory are protected and shared and that traffic inside the box and to the outside world is coordinated and prioritized.

The original Mac didn't have such protection, partly because Jobs insisted that the Mac have the smallest possible system software footprint and RAM size (visiting from France, I was present when team members pleaded with Jobs to let the

Mac have 128K RAM as opposed to the 64K he originally dictated). I'll venture that Jobs' decision to 'go light' with the Mac resulted from the Lisa. The Lisa had a proper, home-grown multitasking OS far ahead of its time. It was also ahead of the day's hardware, and so both sluggish and buggy.

A lack of a kernel makes the footprint smaller and can make some processes feel responsive, but at a cost. Lacking a traffic cop and nanny, every app runs in "superuser mode", entrusted with access to everything inside the machine, including the code and memory of other apps. Every app needs to tread lightly; programmers must ensure they don't wander outside the lines and into another app's territory, apps also must play nice and pass the CPU to the next app when their turn is over, and everybody must trust that other apps are similarly careful and polite.

It didn't always work well. Old timers recall, none too fondly, the dreaded icons announcing a crash (Image #9):

Image #9

This gave us a clear goal: Write an OS kernel and slip it between the Mac's Motorola 68K processor and the current Mac OS.

Simple, no? No.

We didn't have time to write such a complex piece of code, so we looked at companies such as Hunter & Ready, who would license a kernel. Problem almost solved.

But then we realized that the task of delicately lifting the existing Mac software base, disconnecting, and reconnecting blood vessels and nerves, too, would take time beyond our budget; we were concerned that we'd lose the patient. (Lest one think that we were too timid or clumsy, Jobs' team of master computer scientists — Avie Tevanian, Bertrand Serlet, Scott Forstall, *et al.* — faced the same dilemma when Apple bought NeXT in 1997. They took four years to create the Unix/NeXTSTEP-based Mac OS X.)

As we got busy with short-term and medium-term fixes for the Mac product line, Apple engineer Sam Holland came up with another, longer-term idea: Let's develop a quad processor of our own for future Apple personal computers, one that would dramatically outpace the evolution of Intel and Motorola chip development, guaranteeing the Mac's place at the industry's pinnacle.

To simulate the microprocessor, codenamed Aquarius, we bought a Cray supercomputer and used AT&T Microelectronics as our development partner. This wasn't the AT&T we now love to loathe as a wireless carrier. It was the technology giant whose Bell research labs gave us Nobel laureates such as Arno Penzias and Steve Chu and a long list of inventions, including the transistor, cellular telephony, the C programming language, the Unix operating system, the original non-blocking telephone switch and many more.

Although the quad processor development work didn't produce direct results, the Aquarius project stands as an early example of Apple's abiding desire to control the future of its hardware — a yearning which would again manifest (successfully) when Jobs bought Palo Alto Semiconductor to develop the Ax series of microprocessors which now power iPhones and iPads — and the Mac. Apple microprocessors, collectively named Apple Silicon, are now widely considered the industry's best in their categories. The AT&T Microelectronics relationship also led to the Newton project and another company's hardware, but that's for a later chapter.

Our fixes worked for the Mac. My adaptation to both Californian culture and corporate politics is another story.

Firing Frankness

My boss asks me what I think of him. HR advises me to tell the truth. I'm fired.

In a Palo Alto restaurant in January 1990, John Sculley and I have what turns out to be our last dinner together as fellow Apple employees, with the VP of HR in attendance. Over dessert Sculley asks what I think of his performance as Apple CEO, whether I trust his decisions, and whether I think he's leading the company in the right direction. Politely but frankly, I tell him.

I recall the moment's emotion: I felt I was performing a good deed, being helpfully clear and honest, hoping to clear the executive suite's unbreathable air. Before the dinner, HR VP Kevin Sullivan, affectionately nicknamed Kevin the Bold and Sullivan the Magnificent for his unassuming soft touch, had encouraged me to forgo the expected courtier's dissembling and, instead, to 'help' Sculley by speaking straight.

As we wait for our cars, Sullivan puts his arm around my shoulder: "Jean-Louis, I'm proud of you." After half a decade in Cupertino, I know what this means: What I have done is irreparable.

Two days later, Kevin the Bold asks me to come to his office. With his eyes welling tears, he breaks it to me gently: "I've been asked to separate you from Apple, effective immediately".

I return to my office, assemble my team of direct reports and tell them the news myself — I want no one else to spin the story. As I take down pictures and assemble my personal objects, the news races through the engineering organization. Soon, engineers are marching outside with placards that read **Jean-Louis Don't Go!**

The demonstration, small and brief as it is, changes the course of what should have been a typical, quick departure. Apple management becomes concerned that some engineers might follow wherever I land. My protest that I wouldn't dream of such a thing is met with disbelief (perhaps I was too sincere again). The terms of my departure are altered: I'm asked to "stay around" until the end of the company's fiscal year in September.

Thus begins a paradoxically pleasant eight months. As a minister without portfolio, I'm occasionally called upon to offer clarification on unimportant issues, but otherwise, I have little to do. Some HR staffers with their own views of my firing are sympathetic; they ask me what the company can do to make my 'stay' more comfortable. Perhaps I'd be interested in courses on Japanese calligraphy and conversation?

A few days later Sculley enters my office and does a double take. Torn and stained newsprint clutters my desk, big brushes and a well of squid ink, a tall Japanese woman speaking slowly and clearly. HR hadn't told Sculley about my new pursuits. Showing off, I make a feeble attempt to write a big 'ten' (heaven) (Image #10) which is supposed to look something like the proper kanji:

Image #10

I convince Sculley that I will not utter a negative word about my firing, the company I still love, or himself. In my book, that's not done — and I'm genuinely sincere, again. He had been my benefactor in the past, and he was still the boss. When the general and his lieutenant disagree too much, the lieutenant must go. Sculley made the right decision.

It took less than three years to go from the high of shipping the Mac II and SE to that fateful dinner.

Looking back at it, the slow dissolution began in the Spring of 1987 when my friend Steve Sakoman, then in charge of Mac hardware development, told me he was leaving Apple.

Why? Politics.

Sakoman was troubled by the meddling of an expanding stable of upper-middle managers and executives and didn't think it would get better as the company continued to grow. Having made his name by engineering a portable computer, the HP 110, at HP's Corvallis division, Sakoman felt penned in and unused at Apple. He wanted to form a new company to develop a tablet featuring handwriting recognition. I should have given him a pep talk and pointed out that Apple was in great shape with many interesting projects ahead. Without thinking I asked if he needed a CEO.

After discussions with a potential sponsor Sakoman stayed at Apple. We gave him his own organization reporting to me, and a dedicated building on Cupertino's Bubb Road. The Newton project was born.

My responsibilities were expanded, a fancy 'Senior' was added to my title, and we developed the Macs IIcx/ci and the well-rounded Mac SE/30 with a much quieter fan and speedier 68030 processor.

Another project, the Mac Portable, wasn't as successful. Convinced it would be too big and heavy; I wanted to bench the project lead and ask Sony to partner with us to develop a much smaller portable Mac. It seemed like a perfect marriage: Sony was already a well-liked Apple partner; and their talent for miniaturization, aesthetics and manufacturing were indisputable. I got strong pushback on my proposals, complete with accusations of being anti-American. I lost my nerve and capitulated.

In September 1989 I introduced the Mac Portable, assembling it on stage to friendly applause. It was a fun moment, but real-world customers didn't flock to the machine. I got the last laugh though. In October 1991 the PowerBook 100[44], the "tenth greatest PC of all time", was released — designed with and manufactured by Sony.

All along political turmoil continued. In 1988, HR told me that to move up I would have to leave my comfortable

---

[44] https://en.wikipedia.org/wiki/PowerBook_100

engineering building, cross De Anza Blvd, and join the other top execs in a suite of offices surrounding Sculley's.

I still recall my dread: I have no delusions about my lack of courtier's finesse — that was the beginning of the end of my Apple days.

But it would take a while. Just for crossing the street I was rewarded with an even fancier President title (of Apple Products) and added Manufacturing and Product Marketing to my portfolio. The Manufacturing part was especially exciting: I couldn't wait to head over to the factory to 'work' on the production line — where I immediately embarrassed myself by puncturing a loudspeaker while affixing a part to the Mac's front bezel.

Compared to what I saw in Japan and Europe, our factory was primitive and inefficient. And I was dirty, as I found out when I swept under a conveyor at the end of my shift. Some of the other execs question the value of working a few days on the line: How much could I really learn there? "Certainly more than if I hadn't", was my defiant answer.

During 1989 Apple revenue didn't grow as expected. My advice: Raise prices, a 'priceless' piece of wisdom that does nothing for the company.

By this time, I was really on the ropes politically. Proximity to the executives has proven to be the diplomatic disaster I anticipated; my "raise prices" advice was openly scorned; my behavior was considered strange, almost embarrassing. So, imagine my surprise when I got the highest exec bonus for the fiscal year ending in September. I felt vindicated, but the bonus was actually just a *"cadeau de rupture"*, a breakup gift. The next January, Sculley invited me to dinner in Palo Alto.

During my calligraphy hiatus I briefly contemplated an offer to be moved back to France, perhaps as the head of Apple

Europe, an arrangement that appeared less spectacular than being fired. But after a rainy Sunday afternoon spent reading *Barbarians at the Gate;* and an animated evening dinner with a group of French expats, including Philippe Kahn and Eric Benhamou, I realized: "This is where I want to be and what I want to do." Two days later I called Steve Sakoman.

VICI (ALMOST): THE BE YEARS

New Beginnings

Job Offers Before Be

Surveying the field enabled me to figure out what I really wanted to do after a decade at Apple.

As described in the Firing Frankness chapter, my exit arrangement with Apple involved staying another six months as a "minister without portfolio." While I pondered my next move, I got a pair of phone calls from Steve Jobs. In the first conversation he asked how it felt to be fired, a smirking question I deserved given my role in his own dismissal. A few days later, another call on a peaceful Sunday morning. This time Jobs imperiously asked what I was doing. I yielded to the temptation to tweak him a little bit and described my station, in bed, my state of clothing, none, and my occupation, reading the Sunday *NY Times*. Ignoring my response, Steve offered to take a walk and talk, a standard *modus operandi* of his, to explore how "we could do great things together". I declined. As discussed before, I knew I didn't have the emotional strength to work for the charismatic and very controlling NeXT founder. (That was an honest but superficial explanation. It would take a while to figure out where I really stood.)

Other opportunities manifested themselves.

One of them was an offer to run AT&T's UNIX Systems Labs. The overture led to a series of interviews on the East Coast, culminating with lunch with AT&T Chairman Bob Allen and a Bell Labs visit guided by Nobel Prize winner Arno Penzias. This was exciting and flattering but I wasn't sure I'd fit in with that very white, starchy culture. (I asked a Jewish executive what he thought of Chairman Allen's presence on the Board of Trustees of a nearby golf club that did not accept Jewish members and got a shrug in response.) Finally, when AT&T's über-HR executive formally offered me the job, I had to inquire: Pointing to my diamond earring, I asked why AT&T would hire someone like me. 'Because we want you to kick Bill Gates' behind.' I knew AT&T harbored negative feelings towards Bill Gates, whose 'impure' software was usurping Unix's ordained place. I said I understood and would seriously consider this great offer. And declined a few days later.

I also passed on taking the helm of Commodore, a company with a rich personal computer history[45]. Actually, Commodore investors first offered me the job in 1987, after the Mac II launch boosted my reputation, but I had too much fun at Apple, where we would soon start the Newton project.

Commodore chairman Irving Gould saw a better opportunity when I got fired, and we met for lunch during the summer of 1990 at the Four Seasons restaurant in New York City. I was interested because, among legacy products such as the PET, VIC-20 and C64, Commodore had a more modern machine called the Amiga which deserved the kind of treatment we applied to the 1984 Mac. In that meeting Gould helpfully confirmed everything negative I had heard about him and made my decision easy. But as we'll see later at Be, I remembered the Amiga.

Back in the Valley a headhunter called, telling me a Unix server company needed a Chief Operating Officer, with a wink-wink suggestion I would soon become CEO because the founder wanted to retire. When we met the incumbent's

---

[45] https://www.commodore.ca/commodore-history/the-crazy-story-of-how-irving-gould-ended-up-owning-and-killing-commodore/

first question was about the size of my self-image: "I heard you have a big ego. That could be a problem here because you see, I don't have a big ego." I smiled and responded with a phrase of Cupertino-acquired California-speak, while I looked at his office appointment: virile Western bronzes and paintings, massive mahogany furniture, leather chairs. Another easy decision.

I went to Colorado to discuss another CEO opportunity: running Solbourne, a SPARC[46] clone company near Boulder. SPARC was Sun Microsystem's processor architecture. The company looked attractive, supported financially and technically by Japanese giant Matsushita, but a closer look told me they had too little a chance of succeeding against Sun, the main SPARC player.

Back in Cupertino I had a visit from investors offering the delicate task of being Jaron Lanier's[47] boss as CEO of his VPL Research Company. Widely recognized as one of the creators of Virtual Reality, an impressive polymath, Jaron[48] only reluctantly accepted a 'businessman' overseeing his work. Impressed by the depth and breadth of his vision, I felt his work was more research than product development and declined. Three decades later, rightfully calling himself a Renaissance Man, Lanier is the 'Octopus' (Office of the Chief Technology Officer Prime Unifying Scientist) at Microsoft.

I also enjoyed a brief stint as a cashier at Fry's[49]. Fry's Electronics once was Valley geeks' bazaar. Everything you needed for your PC configuration or home automation project was there — along with books, appliances, dried

---

[46] https://en.wikipedia.org/wiki/SPARC

[47] https://en.wikipedia.org/wiki/Jaron_Lanier

[48] http://www.jaronlanier.com/general.html

[49] https://en.wikipedia.org/wiki/Fry%27s_Electronics

ramen and even ties for your next interview. My interest in electronics retail, born during my HP and Apple France days, led to a connection with CEO John Fry. We discovered a shared interest in mathematics, and John proudly opened a vault where he stored manuscripts of famous scientists such as Georg Ohm. John's interest brought Andrew Wiles[50], the famous British mathematician who finally proved Fermat's Last Theorem[51], to a Stanford conference and launched the American Institute of Mathematics, housed in a replica of Granada's Alhambra[52]. Curious, I asked John if I could work at Fry's in my now abundant free time. He chuckled and accepted, setting me up as a cashier for two weeks. It was more illuminating than my passage on Apple's manufacturing floor in nearby Fremont. Fry's organization was disciplined, with detailed processes that went as far as how to array twenty-dollar bills. While making change or processing a credit card charge, I made a habit of asking the customer what they intended to do with their purchase, with questions such as "remodeling your PC." Not everyone was forthcoming, and I got strange looks from the occasional Apple employee who wondered what I was doing behind a Fry's register. I feebly joked I needed to make ends meet, to which a perspicacious individual replied they'd seen my very German car in the parking lot, recognizable for its "My other car is a modem" bumper sticker. My Fry's experience led to an offer to start Fry's stores in France. I was flattered but declined because I had decided to stay in the Valley. That story ends on a sad note. Fry's Electronics is no more — the stores, once favorite destinations for many of us, are no

---

[50] https://en.wikipedia.org/wiki/Andrew_Wiles

[51] https://en.wikipedia.org/wiki/Fermat%27s_Last_Theorem

[52] https://en.wikipedia.org/wiki/Alhambra

more, swallowed by Amazon. And the Alhambra replica won't be built.

The most painful turn down was the offer to run Stanford CCRMA[53] (Center for Computer Research in Music and Acoustics). There Director John Chowning introduced me to former Bell Labs researcher Max Mathews, who gave me a demo of his work. Mathews had a better idea than merely storing a musical score as an array of notes (pitch and duration): he added human 'breathing' to the dry computer rendition by giving the operator a wireless baton. The device worked in two dimensions: vertically (up/down for volume) and horizontally (forward and back for pace). The grand demo involved Max conducting the computer rendering a Mozart aria, along with his opera singer wife. This was all very seductive, as were the facilities at Stanford's Knoll and my offered office overlooking the campus. Again, I asked: Why me? I'm neither a musician nor a research scientist. That's when references came up to Apple France's involvement with a similar institution called IRCAM[54], in Paris' Centre Pompidou, headed by famed composer and director Pierre Boulez. True, we had donated a few computers, and I was treated to a Jupiterian demo of IRCAM's 4X computer synthesizer, where my digitized voice became an instrument in a Beethoven symphony.

The truth dawned on me later, at a tasting at the Thomas Fogarty winery on top of the hill overlooking Silicon Valley. The event honored Yamaha, a significant CCRMA donor. I was introduced to company executives as someone recruited to head the institution — and finally saw my future. As head of CCRMA, I'd host a neverending succession of such events. My CCRMA *œuvre* would be the music of cash

---

[53] https://ccrma.stanford.edu

[54] https://www.npr.org/templates/story/story.php?storyId=97002999

registers, I'd be chief fundraiser. I decided that if I had to be in perpetual fundraising mode, I might as well do it for an enterprise of my own.

A small coda, just for fun. It regards the diamond earring I mentioned at AT&T and Bill Gates' quick wits.

Later in 1990 at an industry conference, I donated the JLG earring for a charity auction. Bill Gates won and immediately came to me demanding to get his prize on the spot. I answered that I would be delighted to oblige on one condition: I would install it myself. To which Bill wittily replied that wouldn't work because, guessing the part of his body I had in mind, he knew well no one would see the expensive ring he'd just won.

My Debt to Efi Arazi

How my sometimes-abrasive ways earned me a friend and sustenance for my family, and the early years of Be.

While contemplating life after Apple, an earlier and politically charged engineering effort played a helpful role in my transition. In surprising ways, the company's TrueType project paved the way for a source of financing and advice for my next venture.

This tale is twisted. It involves the charismatic Efi (Efraim) Arazi, founder of the eponymous EFI (Electronics For Imaging); creative engineers and tremulous execs at Apple; a feud with Adobe; Fred Adler, the crusty financier behind Data General and EFI; and yours truly contemplating life after Apple. As an addendum, I'll have some fun with the tired but inextinguishable myth of Apple's Walled Garden.

It was a spring afternoon in 1990. I'd been fired from Apple a few months earlier, yet was still in my office in Cupertino, getting ready for a calligraphy lesson while I served out my severance time. The door opened but it was not my sensei; instead Efi Arazi walked in. Having charmed his way past my assistant, the legendarily uncajolable Ms. VV, Arazi picked up a sheet of newsprint covered with attempts at a particularly difficult Japanese character, examined it, and then bowed and said, "I'd be honored if you would join the EFI board."

"Why me?"

He turned the newsprint around and waved at my handiwork. "Because of your diligent brushstrokes".

I chuckled politely at a joke I didn't understand, but he was not being entirely facetious.

The explanation requires us to back up a few years.

In 1984 Adobe introduced PostScript and revolutionized desktop publishing. One of the most important aspects of PostScript is its typeface system, known as Type 1 fonts. Unlike bitmap fonts such as the celebrated Chicago typeface, for which separate bitmaps had to be created and stored for each character at each size, Type 1 characters contain drawing instructions to be rendered at any scale.

Apple popularized PostScript by adding it to the original LaserWriter, but Adobe's creation's flexibility and indisputable quality came at a price: about $30 for a basic font package. This relatively big slice cut into the Mac's thin margin, but we had no choice until an Apple engineer named Gifford Calenda proposed his own "mathematical" typeface system. The effort cost a couple years' salaries for a small team — a pittance compared to the Adobe tax. After checking with Apple's intellectual property attorneys ("no problem"); I told Calenda and his spouse Sheila Brady, product manager for the effort, to get to work. Here we recognize an enduring and immensely valuable trait of Apple culture: controlling key parts of its product, *a.k.a.* the 'stack' of hardware and software layers that ultimately undergirds applications — the user experience or UX.

Apple was known as a ship that leaked from the top. Adobe execs quickly heard our plans and suggested I be fired for incompetence: What does Gassée know about the intricacies and aesthetics of a typeface system? Everyone knows Adobe is The Reference in this domain.

Apple execs were shaken — Adobe's prestige was impenetrable and its role in helping us start the desktop publishing revolution was critical. One could even say Apple wouldn't have survived without Adobe's help. Fortunately prototypes of our invention came up quickly and were convincing. TrueType, as it's ultimately called, was announced in 1989, with an unexpected spin: to ensure wide adoption — and because we don't need to make money, we just want to spend less — we do the unexpected and give Microsoft a free TrueType license. Adobe CEO John Warnock isn't pleased. He gets very emotional and accuses Apple and Microsoft of selling "snake oil," arguing we're purposefully and spitefully torpedoing Adobe's font business.

That brings us back to Efi Arazi, whose company created a sophisticated networked color printing server. He knows EFI will tangle with Adobe over computer imaging issues, and he wants someone on his board who has had success in that fight. Today, more at peace with myself and the world, I smile when considering how that feisty, arguably asshoholic, facet of my behavior — for which I reproached myself at the time — came to be an asset at a crucial time.

In addition to joining EFI's board, Arazi offers to sell me 1% of his company for $175K. Unfortunately, I'm nearly broke. I've just completed an expensive home project in Palo Alto, and my final modest exit settlement from Apple won't materialize until the end of September, some months away. A few days later, Arazi calls back with good news: Fred Adler, his lead investor, asked Arazi to "give" me 1% of EFI as stock options. Adler has a favorable recollection of my days at Data General, a company he bankrolled many years before.

After EFI went public (and the company reached a settlement with Adobe), my options were worth much, much more. This enabled me to feed my family and my new company, Be. In addition to financial generosity, Arazi also gave me advice on my new venture. As we'll see in the next chapter, *"Be: from Concept to Near Death"*, I did not follow it.

While the eponymous company continues to play an important role in multiple facets of color printing, Efi Arazi is no more. Memories of his scintillating intellect, humor and friendly grace live on in my heart. I am grateful for his entering my life at a crucial time, and his generosity.

I stayed on EFI's board for 17 instructive years. It was the start of another side of my life after Apple, sitting on boards such as Cray Computer Corporation, Laser Masters, 3Com, Logitech, Palm Source, and CATC. I'm grateful for the diverse experiences, both good and less positive. They all redound to a body of knowledge and empathy I now try to pass on to fledgling entrepreneurs, fully accepting that they, like me, won't always listen.

Apple Years

Credit: Ann E. Yow-Dyson 1988 – Personal archives

Ed Kashi via Getty Images

Be Years

Credit: Eric Sander- 1997/98 – Personal archives

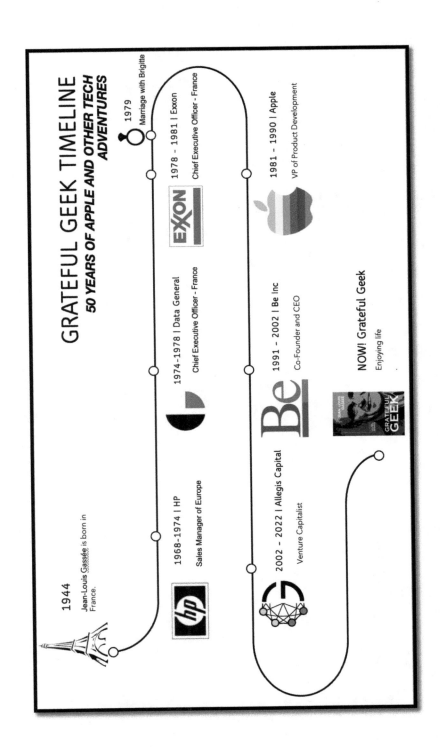

GRATEFUL GEEK TIMELINE
50 YEARS OF APPLE AND OTHER TECH ADVENTURES

1944
Jean-Louis Gassée is born in France.

1968-1974 | HP
Sales Manager of Europe

1974-1978 | Data General
Chief Executive Officer - France

1978 - 1981 | Exxon
Chief Executive Officer - France

1979
Marriage with Brigitte

1981 - 1990 | Apple
VP of Product Development

1991 - 2002 | Be Inc
Co-Founder and CEO

2002 - 2022 | Allegis Capital
Venture Capitalist

NOW! Grateful Geek
Enjoying life

ENTREPRENEUR'S LIFE

Be: from Concept to Near Death

Be, the company Steve Sakoman and I founded after leaving Apple, had a great product idea — or so we thought, only to discover that a bad choice of partner would almost destroy the company.

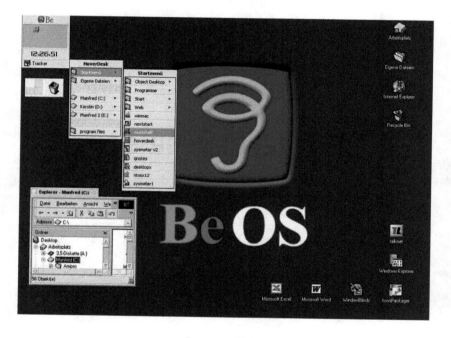

Image #11

Going back to February 1990, shortly after my exit from Apple was arranged, my friend and colleague Steve Sakoman came to my office and announced his intention to leave the company. I was surprised; Sakoman was well-regarded, he ran a separate organization developing an exciting new product, the original Newton tablet. (Oddly, the Newton Wikipedia article omits the first full-size tablet phase of the project.). But no. Despite his promising future at Apple, Sakoman wanted to leave because he had doubts about the company's direction (as described in chapter *Firing Frankness*), and concerns about Newton's handwriting recognition, a key feature of the device.

When my half-hearted attempts to convince him to stay at Apple foundered, I did the only sensible thing: I offered to start a company together. While I didn't have a precise idea about a product, I was convinced we'd make a good team, as we had already. That proved correct. I also told Sakoman I'd fund the project, perhaps with friends' help to avoid falling prey to Vulture Capitalists. This proved severely misguided and almost lethal.

For a while we spoke in circles, failing to find the core of an idea. Then in July, as I was vacationing in France, we discussed over the phone our frustrations with the complicated layers of hardware and software silt the Mac had accumulated. Surely, we could come up with a simpler, cleaner architecture, a more agile personal computer — but was there room for a third product besides the Mac and Intel PCs?

In our ensuing conversations, the Commodore Amiga became our "reference platform," an example of the good, the bad, and the promising. While the Amiga provided interesting multimedia features, we were less impressed by its overly complicated architecture and cobbled-together implementation. But the most important aspect to us was that it sold in attractively large numbers, proving there was a market for a third way.

Weeks later, Sakoman presented a basic sketch of a simple, powerful multimedia machine. During Newton development and our work on the Aquarius quad processor, we developed a relationship with AT&T's Microelectronics division in

Allentown, PA. The original Newton project used two Hobbit processors: simple and inexpensive RISC[55] devices. AT&T also made DSPs (Digital Signal Processors), simplified, capricious, but fast chips that process the digital samples that handle images, video, sound, telephony, fax transmissions, and the like on computers. Sakoman's idea was to build a multimedia computer featuring two Hobbits and three DSPs (one each for sound, video, and fax/telephony) on a simple homegrown bus.

Sakoman's lab was in his Scotts Valley home, more precisely in his "clear room" (as opposed to the dark room where he developed negatives, color no less). There he delicately retouched prints from his 8x10" field camera. Soon I procured expensive (for us) lab equipment, such as a fast Tektronix oscilloscope and an HP Logic Analyzer. For that I had to call HP CEO John Young's office; we weren't yet incorporated as a business, so HP's salespeople wouldn't ship to Sakoman's private home.

In a matter of weeks Sakoman's hardware prototype produced a heartbeat. Now, all we needed was a software team — built around Erich Ringewald, an Apple alumnus; Cyril Meurillon, a French engineering student I failed to convince to drop out of the prestigious Sup Télécom school (he wrote key parts of the OS kernel in his Paris dorm); and Benoît Schillings, a blindingly fast prototyper from Belgium. A little later we recruited more Apple alumni: engineers Bob Herold and Steve Horowitz, and attorney Cory Van Arsdale as business manager.

I had what I thought was a great name for our now-incorporated company: United Technoids, easily abbreviated as UT. This was both cute and useless, as there

---

[55] https://en.wikipedia.org/wiki/Reduced_instruction_set_computer

was already a huge conglomerate called United Technologies.

During one of our daily phone calls, Sakoman told me he and his family had hunted for a name the night before by crawling through a dictionary. They'd gotten as far as "B" — I thought he meant the letter B. No, "Be" as in "To Be" (Image #11). Simple, resonant, romantic, even — a call to action. I loved it.

At home I pulled the OED Second Edition and walked through the long (25 page) procession of the verb be's meaning through the ages; including its Sanskrit, Saxon, and Latinate roots (Be, Was, Is). I immediately jumped into MacDraw to design a simple Times Roman Light Condensed logo, manually kerning the "e" to ensure it looked right next to the "B". We were in business — after an amicable legal settlement with a company called Better Education.

Early demos impressed visitors to Sakoman's clear room, enough that an early investor friend wrote a second check on the trunk of his rental car at the end of his visit. (See the next chapter on the Don'ts of my dangerously naive fundraising efforts. In short: Skip Friends and Family, go for cynical Deep Pocketed Pros.)

Then disaster struck.

We should have seen it coming. At the beginning AT&T Microelectronics was very supportive. They knew us, they worked closely with Sakoman on the Newton project, they exhibited tepid optimism for my ability to generate publicity for Be's multimedia machine and their processors. But as time passed, we experienced difficulties getting a clearer product roadmap, and in 1993 we got the news: AT&T was giving up on the Hobbit.

Gestation interrupted.

This hurt Sakoman badly. I tried to convince him this was a mere bump in the road. I used a literary analogy: the author's computer and backup burn in a freak electric fire, the manuscript is destroyed. "I feel your pain", says the publisher, "We'll get you another advance, you have the book in you, you'll even have the opportunity to fix some problems you were telling me about; success still is just around the corner."

Easy to say but engineering design is a human creative activity, rational only on its thin surface. Sakoman was distraught and couldn't bring himself to start over. He left. But he returned a few years later.

I lost a colleague, but not a friend.

A few months earlier I almost lost my life. I tore the inner lining of my left carotid artery and experienced a stroke as a result. I didn't understand what had happened, but lost speech for a few minutes and was sitting at home when my wife discovered me, dragged me to Stanford's ER and probably saved my life. Fortunately, the only sequela seems to be a numb patch of skin on my right index finger.

When we 'lost' AT&T and the Hobbit, my friend Efi Arazi told me to shut Be down and apologize to investors, who would have to understand the impact of losing a crucial supplier, and then come up with a new idea. I didn't follow his advice.

Be Fundraising Misadventures

Sneering at venture investors, calling them names such as Vulture Capitalists, is a long-standing Valley pub topic. I had to learn that cynical pros are less dangerous than a naive entrepreneur taking money from friends.

Summer 1991. Be, the company Steve Sakoman and I started the year before, was ramping up, but I was merely the CEO writing checks rather than code, so I took a short family vacation to Arcachon in southwest France.

I was daydreaming past seaside shopfronts when something caught my eye. I backed up, peered in the antique store window, and there they were: two ceramic pigs, each about two-feet tall, dressed as butchers (Image #12). The perfect avatars of my disdain for the Venture Capital profession. Unsubtly I christened them Victor and Charles:

Image #12

After all these years — and even after becoming a member of the VC brotherhood — they're still on my desk. If only they could talk.

I wanted to keep Be out of the vulture capitalists' talons, so to fund the company in its early years, I put my own money into the venture. That was the first of a series of fundraising mistakes.

My first excuse is that I used to be French, from a distrustful culture. In France the entrepreneur was often viewed as a mountebank keen to take advantage of investors' funds. To prove their good faith, French entrepreneurs were expected to put their own money on the line. It's better now.

So, I thought I was doing the right thing, but I found out professional investors in the US are suspicious of self-funding. They prefer a clean division of labor: The entrepreneur provides the idea, psychic energy, and leadership; the pros supply financial fuel — and advice, often disregarded.

When my personal coffers ran low, I accepted investment money from friends and business acquaintances. But taking money from friends can be lethal: the money always runs out, friends don't have professional VCs' deep pockets, and the company becomes vulnerable.

But first the Law of Professional Venture Investing.

A pro invests as much money, as many times, for as long as required for the situation to attain Clarity — which means one of two outcomes: either Dead or Liquid. When a professional VC declines an invitation to invest in a follow-on round, it means they reached Clarity: the pro who declines to invest has decided the company is Dead and their initial stake is worthless. Write it off and move on, no tears, no recriminations.

But if another pro keeps the company alive by staking more money, they do so at elevated risk, and so they pay less — often much less — for the new shares they buy. As a result, original investors are diluted by the new money: a 10% share in a previous round might become worth 2% or less.

This might seem counterintuitive and unfair: Shouldn't the original investment, the one that took a risk on an unknown, be worth more than the newcomers? This is another part of the Law: If a company needs additional rounds of funding because it's not growing as well or as fast as anticipated, it is now seen as riskier. As a reward for taking a higher risk on a venture now foundering, new investments get more ownership per dollar. (None of this applies to follow-on rounds for a happily growing entity.)

Back to Be.

A good friend introduced me to the COO of Crédit Lyonnais, France's largest bank. The meeting went very well, and the senior exec told the bank's venture arm to invest in Be.

I was thrilled. We had deep friendly pockets, and the deal struck a high valuation I was proud of for a while. I didn't realize it would discourage other potential investors, who couldn't see a way to a significantly higher multiple of the Crédit Lyonnais terms. Another misstep.

But wait, there's more.

In 1993 Crédit Lyonnais went belly up, right when we needed more money to move beyond the end of Sakoman's Hobbit design. Crédit Lyonnais couldn't put more money into the company. Our lead investor effectively died and almost brought us too.

Seeing the bottom of the cash drawer, I again went on a fundraising campaign, this time in a weakened position. Thankfully friends extended help; my spouse, Brigitte, felt we couldn't let the product die, and agreed to stake more family money. Desperate to keep the company going, I even gave a presentation to a French investor three days after I was released from Stanford Hospital, unshaven and with metal clips on my neck where the neurosurgeon had patched my carotid artery. We kept the company alive, barely, with a motley crew of backers.

Then an Apple alum saved us.

In the summer of 1995, I had lunch at Menlo Park's Cafe Borrone with Lia Lorenzano (now Lorenzano-Kennett), the logistics director of a now deceased industry conference called Agenda. Although I protested that Be wasn't ready for the limelight, she wouldn't let me go until she convinced me to present at the next Agenda a couple months later.

I returned to the Be offices across the street and confessed my just-committed sin to the horrified engineers. "We're not ready!" It didn't matter. We were on the Agenda.

I flew down to Scottsdale a few days before the conference to prepare, but I was blocked, paralyzed with stress. Fortunately, my good friend Jean Calmon, the Apple France Sales Manager who joined Be to help us with the European side of our business, put together a slide presentation. Be employees and Brigitte flew in right before the conference started. On the day Steve Horowitz, one of our earliest engineers, gave a masterful demo while I stumbled through Jean's slides. (Later Horowitz would give the first video demo of Android.)

The response was astonishing: We got a standing ovation, only the second one in the conference's nine years. I was left close to tears and almost speechless. The multimedia performance we had dreamed of when starting Be stunned the crowd. We got good press, including a nice NY Times article. And, at long last, we felt actual interest from the Silicon Valley professional investors in giving us a new lease on life.

The lead venture investor in the next round was August Capital's David Marquardt, the only VC who invested in Microsoft before its 1986 IPO. He was also a Microsoft board member, a position that made life 'interesting' from day one of our relationship.

The friends and acquaintances who had invested in Be were thrilled to see support from Valley VCs, but then they saw the terms. I had the humiliating task of telling them that Marquardt wanted a 'clean' capital investment, which would dilute previous ownership by a factor of 300! If they refused to sign off on the terms, there would be no more money and Be would die. (I was also told the money I had lent the company 'stayed', meaning it was gone, along with the other investors' money.)

To add to the hurt feelings, the Be team — including yours truly — would receive fresh, undiluted stock options. This was another application of the Law: We kept investing (ourselves) in the venture and would be treated as "new money". But it left a sour taste in the mouths of some earlier investors, including Crédit Lyonnais' venture arm, which should have known better: "We supported you, and now we're left with nothing while you go off with glorious new money, fresh stock options, and media attention!"

Everyone reluctantly but helpfully signed off, but any of them could have killed the deal — and Be. To this day, I'm grateful to them, and embarrassed for the consequences of my naiveté.

Victor and Charles might be difficult, greedy, and whatever other negatives you might want to throw at them, but they're deep-pocketed pros who serve a purpose. And there are many to choose from in the Valley. The trick is to choose wisely.

The next chapter considers the last part of Be's life.

From One Ice Floe to the Next

Armed with solid funds from a prominent Valley Venture Capital syndicate, we set out to build and sell the second version of Be's hardware, the BeBox. We soon found that we needed a bigger hardware platform.

When the ice floe begins to melt, the small enterprising tribe looks around, jumps to a larger slab of ice, and then on to another. In retrospect this is what Be's decade-long adventure looks like as well.

As described in the previous chapter, Be jumped from family money to the friendly investors at Crédit Lyonnais, France's largest bank. When Crédit Lyonnais went belly up, we tried a motley crew of helpful but not immensely affluent backers. Finally (hopefully), we moved on to the sound capital structure provided by a syndicate of deep-pocketed Valley investors led by Dave Marquardt's August Capital.

Over in hardware we were performing a similar dance. After AT&T Microelectronics abruptly stopped making the Hobbit microprocessor at the heart of Steve Sakoman's dual-processor design, we jumped to Joseph "Joe" Palmer's dual-PowerPC 603 BeBox.

Swapping the Hobbits for two PowerPC 603s proved more difficult than a 'mere' operating system port, a difficult task in its own right. The 603 wasn't designed to support the dual-processor operation that provided Be's distinguishing performance boost to media applications. Palmer had to design special hardware that, among other challenges, handled delicate caching operations. This resulted in the BeBox (Image #13):

Image #13

With the money just raised from a group of VC firms headed by Marquardt, we were finally gaining the ability to build, sell, and support our system.

The relationship with Marquardt was unusual in several ways. He showed great interest in our project, calling me almost daily to check on our progress and talk shop. As a Microsoft board member, Marquardt had a rich trove of sometimes spicy anecdotes about company key players, which he generously shared for my amusement and education. I must add that he never related confidential business information, which allayed my concerns about bidirectional sharing.

With funding in place, we found an electronics manufacturing contractor to build BeBoxes and started a developer program for the Be operating system, BeOS. *Monday Note*[56] regulars who read last year's "Ten Years of Monday Notes"[57] will recall we started a weekly newsletter[58] for Be developers in late 1995. Coincidentally Microsoft launched a similar newsletter shortly thereafter.

Well-liked by its audience of code writers, the Be Newsletter also covered Sales, Marketing, and PR topics. If you're a techie interested in history, over twenty years later those weekly missives still read well. (The Newsletter's editor-in-chief, Doug Fulton, performed merciful surgery on my Monday Notes.)

The $1,600 66MHz and $2,995 133MHz BeBoxes were met with what is gingerly called succès d'estime (nice reviews), a euphemistic way of saying that sales didn't meet the level required to make the BeBox a self-sustaining business. Developers loved it, but normal end-users didn't feel the BeBox to be a safe enough bet. The ice floe was shrinking again.

Over in Cupertino, meanwhile, these were not the best of times. Michael Spindler, who succeeded John Sculley in 1993, opened the market to PowerPC Mac clones, hoping to convince customers and developers that Apple was broadening the reach of its architecture.

That was our opportunity, a new ice floe for BeOS.

I can't precisely recall who got the idea for the jump — it might have been Bob Herold, a team member from the early

---

[56] A weekly column, now more than 14 years old, published with my friend Frederic Filloux, mostly focused on the Silicon Valley scene: http://www.mondaynote.com

[57] https://mondaynote.com/ten-years-of-monday-notes-645734e25cdd

[58] The Be Newsletter Article Index is preserved at http://testou.free.fr/www.beatjapan.org/mirror/www.be.com/aboutbe/benewsletter/article_index.html

days — but with help from insiders we gained access to Apple's firmware and hardware connections. In short order we made the code changes and saw our operating system run on Mac-compatible machines such as those sold by Power Computing. We were in time to be announced at a Macworld Expo, where we booked a booth. Our announcement raised more questions than we could answer but created a fresh burst of publicity.

Seeing our software run on Mac-compatible hardware (and some Macs on the side) got tongues wagging. Why didn't Apple replace the aging — some people used worse words — MacOS with a BeOS core? Informal conversations started during the summer of 1996 and led to serious negotiations, including execs and bankers.

We know the end of the story: instead of acquiring Be, Gil Amelio, who replaced Spindler in early 1996, bought NeXT and its NeXTSTEP operating system instead, with Steve Jobs returning to Apple as an adviser.

One version of the sory is that Be and its investors wanted too high a price, $400M, while Apple's first offer was in the $120M range. Intemperate words were uttered by our side, namely yours truly and a lead investor, and the gap couldn't be bridged. I was mortified after failing to lead our team and work onto firm ground.

Later I heard another version of the story from a NeXT and Apple insider. By 1996 Jobs had given up on NeXTSTEP and focused the company on WebObjects, a set of tools and building blocks aimed at designing and implementing high-performance, feature-rich websites. But the noise from the Apple-Be conversations attracted the attention of Avie Tevanian, VP of software at NeXT, who urged Jobs to dust off NeXTSTEP and whip up a demo. The master persuader convinced Gil Amelio to pay the price Apple didn't want to pay for Be: $429M for NeXT.

Valley wags insist it was a reverse acquisition. After he returned to the company he co-founded, Jobs and a cadre of supporters (we could call them "Deep Apple") engineered a board coup and ousted Amelio. To be fair, Amelio deserves credit for stabilizing the patient (Apple's finances) by bringing in a highly respected CFO, Fred Anderson, whom Jobs eagerly embraced. And we owe Amelio a debt of gratitude for bringing Jobs back to Apple and inaugurating the Apple 2.0 era, a turnaround the likes of which our industry has never seen, before or after.

For Be the money was running out again, and we looked for yet another ice floe, the topic of our next chapter.

Be's Last Ice Floe...

BeOS and its creators kept jumping to new ice floes, until the adventure ended with a 1999 IPO and a low-price sale to Palm after the 2000 burst of the Internet bubble.

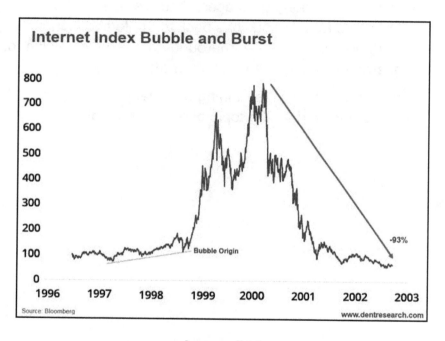

Image #14

In the last chapter, we saw Gil Amelio's felicitous move of historic proportions. He left Be at the altar and brought Steve Jobs back to Apple instead. For his reward Amelio was knocked off his CEO throne.

Apple's decision to go with NeXT and Jobs was doubly perilous for us. Not only would BeOS not undergird the next MacOS, but Jobs immediately walked third-party Mac hardware makers to their graves. No more Mac clones for BeOS to run on. With tepid BeBox sales and no future on the Mac, Valley VCs weren't keen on another round of funding — and the 1995 round was running out (Image #14).

Our little tribe needed a new ice floe.

With a fine sense of the retroactively obvious, we noticed the rich and woolly universe of Intel-powered devices that ran variations of DOS, Windows, Linux, and Unix operating systems. In short order we had a BeOS demo running on Intel hardware and took it on the road. One Intel research exec exclaimed he didn't know a PC "could do this" — offering multimedia performance impossible on Windows. Intel was to anchor our next funding round.

Another stroke of luck occurred in 1997 when François Pinault, founder of financial giant Artemis (and a noted art collector), showed up at our Menlo Park office and inquired about investment opportunities in the Valley. I launched into a 'learned' and, as it turned out, embarrassing exposé on the Valley's reckless and unsophisticated moves, warning him about pitfalls that would horrify French investors. Mr. Pinault politely interrupted my lecture to explain that, acting on a tip from his more technically inclined son François-Henri, he was asking about investing in Be.

With a bit of fresh money, we set out to find buyers for BeOS licenses. Our idea was to get PC makers to offer BeOS as an alternative to the standard Windows OS in a dual-boot configuration. At startup the PC would offer to launch Windows or BeOS. PC users would feel safe that Windows would always be there, but also exposed to the superior performance which excited Intel execs.

There was, however, a fly in the dual-boot lubricant. While the Windows license allowed PC makers to offer a dual-boot tool, it had to be Microsoft's dual-boot launcher — which only recognized Microsoft operating systems. We were stymied.

Later, after Be's assets — although not the company itself — were sold to Palm, we sued Microsoft and reached a settlement in the $20M range. With questionable humor, I called it enough money to put fresh tires on my (imaginary) wheelchair.

Yet again we needed a new ice floe. This time we jumped to the newly *en vogue* Internet Appliance concept. We presented BeOS as small and agile, yet powerful enough to enable dedicated Internet application devices. At the time Comdex, the since-defunct computer trade show, had a pavilion dedicated to Internet appliances, where I once saw Bill Gates checking out the new interlopers.

With the money running out again in the first quarter of 1999, Be CFO Wes Saia took me aside: let's take the company public on the strength of our Internet Appliance platform. But, I protested, we haven't finished the platform yet, and we have little to show in terms of revenues and profit. It didn't seem to matter. Saia had met with some bankers he knew from experience running previous IPOs. They agreed that Be was a strong candidate for a 'concept' IPO, instead of one based on a more conventional business. Please recall that the stock market was white hot, and bankers were hungry for any tech company they could take public.

So, we prepared an IPO roadshow, complete with a demo Internet Appliance carefully tended by George Wong, a Be engineer who also was a dentist. His delicate 'digital' skill came in handy as we took our fragile prototype to both US coasts and Europe.

During the road show I was forced to learn brevity. When our bankers couldn't convince me to cut back my opening act, they simply moved me to the end of the show, after the Marketing

VP, the CFO, and the demo. As each pitch was limited to 25 minutes — there were other companies in the queue behind us — I had no choice but to select my few words carefully. The forced austerity became the root of Monday Notes such as Three Slides? You're Nuts! OK. How About Seven?[59].

With a few complications and a lower price than we had hoped, we became a publicly traded company by the end of July 1999, with the BEOS ticker symbol.

At the same time, I became peripherally involved in the antitrust lawsuit against Microsoft. As a potential witness I met with David Boies, lead attorney for the US government, and the DOJ's Antitrust Division head, Joel Klein. The latter advised me to knock back a stiff scotch before weighing my decision to give testimony — he thought I was 'dangerous' — meaning unpredictable. On our CFO's advice, I gave our IPO priority.

Right after we went public, I took our extended family (the flesh-and-blood sort) on a different sort of roadshow, up Interstate 5 to Vancouver, BC and back again. The drive got me thinking: Be would not make it. We hadn't raised enough money in our IPO, and I didn't like what I saw in the Internet Appliance market.

Upon returning I told our board we should sell the company right away. My plea fell on deaf ears, I was told we should use the IPO money to seed the market with prototypes and "do a secondary", meaning sell more BEOS shares in the spring, a fairly common maneuver back then.

I had a choice to make. I could resign and sell my shares or ride it out. I owned 10% of the company, and as a side effect of the

---

59

Three Slides Then Shut Up — The Art Of The ... - Monday Note
https://mondaynote.com › three-slides-then-shut-up-the...

Three Slides? You're Nuts! OK. How About Seven? | by Jean
...https://mondaynote.com › three-slides-you-re-nuts-ok-h...

Microsoft antitrust situation BEOS market capitalization briefly reached $1B. My sell-out price was looking like serious money.

But if I resigned the actual sale price was very questionable. I decided to stay — I couldn't abandon ship, even over a key disagreement with the board. Seeing my lack of optimism, the board wisely appointed Steve Sakoman, who had come back to help with the Intel and Internet Appliance efforts, as COO.

Then in 2000 the Internet Bubble burst. Dozens of web-based companies folded, and leading companies such as Cisco lost over 80% of their market price.

We were out of floes. In 2001 we sold Be's assets to Palm for $11M, about 90% down from its peak value. After the sale a clash of company cultures caused many Be employees to quickly leave. They are now doing solid technical work at companies such as Google, Facebook, and Apple.

I later traded my seat on the board at 3Com to become Chairman of PalmSource (the software arm spun off from Palm). After making a variety of changes, we sold the company in 2005 to Japan's Access Systems for about $324M.

While I was on the PalmSource board, I learned that Manuel Petit, a former Be employee with a gift for making low-level connections between hardware and software, had offered $800K for a BeOS source code drop, no support expected. Why? An unconfirmed story at the time held that BeOS was being considered for a tablet/smartphone project at Apple. He was rebuffed — Palm wanted $1M.

Today a BeOS-compatible OS is still available, the free and open-source Haiku OS.

THE LATE WISE (AND LEARNING) YEARS

UNDERSTANDING THE MATHSET* OF A GEEK LEADER

*Mathematical Mindset

ENTREPRENEURSHIP AND INVESTING

Victor & Charles: We Meet Again

Ten years as a Be entrepreneur changed my understanding of the world of venture investors.

Readers will recall that in 1991 I brought two almost two feet tall ceramic statues on vacation in France's Arcachon. Costumed in turn of the twentieth century French provincial garb, one was dressed a cattle merchant and the other a butcher. With my less than positive views of Venture Capital investors, I saw them as the kind of money-grubbing and controlling business partners I didn't want aboard my new Be venture; unsubtly I christened them Victor and Charles.

Ten years later, after the final sale of Be to Palm, I initially resisted my mentor Barry Weinman's entreaties to join Allegis Capital, where he held the rank of Managing Director. My stated fear was going of blind from looking at endless successions of PowerPoint presentations. So, for a short while I helped a small network technology company called CATC update its management team. I brought in a talented new CFO named Carmine Napolitano, who became CEO soon after I left and then sold CATC to a larger company in the same technology sector, Finisar.

After this I finally listened to Barry's advice, and entered a new world of experiences which greatly enhanced my understanding of the world of entrepreneurship — and my own personality.

Based on my years at Allegis, I offer an empathetic set of reflections on the challenges entrepreneurs face when raising money and running a startup.

Many ideas, practices I couldn't see inside the entrepreneur's fog of war, emerged and coalesced as I sat in hundreds of presentations made by my ex-brothers in arms. What follows is a distillation of those ideas: dos and don'ts I owe those presenters.

Three Slides Then Shut Up: the Art of the Pitch

The best pitches aren't really pitches. Dumping one's entire body of knowledge on easily bored investors won't help. The best pitch is one that quickly moves from monologue to conversation.

The First 70 Minutes of The Hour.

When in 2002 I finally listened to Barry Weinman, my Gentleman Capitalist mentor, I worried aloud about looking at PowerPoint presentations for the rest of my life. Gentleman that he is, Barry didn't — and didn't need to — remind me of the two-hour investment pitches I had inflicted on folks like him during my early entrepreneur days.

As mentioned earlier, I finally learned to curb my prolix talk during the Be IPO Roadshow in 1999. The investment bankers who helped prepare the show disabused me of my long-winded ways. I was relegated to the cleanup position, following the VP of Marketing, our experienced CFO (three IPOs before ours), and the demo. Putting me last before the hard stop forced concision.

Having joined the VC brotherhood and being on the receiving end of money-seeking tall tales, I can attest that my fear of mental cauterization by PowerPoint was not misplaced. I have found a name for this blight: The First 70 Minutes of The Hour.

The condition occurs when an entrepreneur uses their allotted hour to dump everything he or she knows about his or her business. I reminisce on my own sins: I was anxious, unsure which of the product's many arcane features and benefits might click and terrified that I'd leave something out. My desperation induces a combination of confusion and mental paralysis as the allotted hour ticks past, and I receive non-committal California-speak: Great, Interesting, We'll circle back to you.

This is an unfair caricature, but not far off. Too many presentations concentrate on the speaker's needs instead of the audience's interests. Fortunately, there's a simple remedy: Show three slides and shut up. Say just enough to engage and then move on to a lively conversation: to questions, arguments, and suggestions.

The canonical three slides go like this:

1. *Who we are*: The founding team's résumé — including technical, business, and academic background.

2. *A nice sharp dichotomy*: The world before vs. after us. Show a substantial and practical impact, not just a marginal improvement of something already in place. The more impossible or unthinkable, the better — it will become retroactively obvious once understood. Consider the computer world before and after the mouse.

3. *The Money Pump*. Your business plan. I like the money pump image, of pipes that allow the cash temporarily residing in customers' pockets to flow into the company's coffers — legally, willingly, and repeatedly.

After that, shut up.

The silence will be unbearable. It might help to look down at your shoes, your hands, or something on the conference room table. But the awkward moment won't last more than an interminable 12 to 15 seconds. If you don't get questions, you have your answer: We're not interested.

But if we poke holes in your story, demand explanations, or play devil's advocate we're hooked. You may now dig into the 253 backing slides you have under the table, whip out the market research, competitive analysis, academic studies, and financial projections — now you can casually lay out your roadmap. Show that you're not afraid to think on your feet. You can even gently flatter us that we're the visionaries — you just want to help make that vision a bit clearer.

You're either in or you're out, but you won't have wasted our time or yours.

There are benefits to this approach, even when we don't buy your pitch.

If we've turned you down, you can call us back six months later, reminding us of your 'failed' three-slide presentation and offering to show three new ones. If the first pass was quick and painless, we might ask you back in. You won't get this welcome if you bored us for 70 minutes the first time around.

Moving forward, sharpen the internal characterization of your business. You can't have ten success factors which are equally important. Concentrate on the top-level features in your Before/After slide and leave the "really cool" pet tricks for the ensuing conversation. Remove the branches that blur the picture, but don't hack away at the graphical details in your slides. Edward Tufte, the world's pre-eminent "data visualizer," posited the counterintuitive notion that by adding visual cues we enhance comprehension. (We'll return to Tufte in the postscript.)

And the most important benefit: If you've distilled your presentation into three slides, you won't even need them. The effort will have been so intense that they're now burned into your brain. You can walk into a conference room, ask for a whiteboard and a marker, and impress us with your command of the business by 'extemporaneously' drawing three slides. There will always be time to whip out your laptop, tablet, or big smartphone for the 253 FAQ (*"Foire Aux Questions"* in French) slides.

This is easier said than done. I relate to anxious entrepreneurs who have a hard time sorting through the wonderful ideas brewing inside the garages in their heads. Afflicted with what Buddhists call 'monkey mind', I too have difficulty quieting the noise to 'hear' the most important, reality-changing element of a product/service/business. Only the most gifted and focused (and most delusional) can see the blade's edge with unfailing clarity. The rest of us muddle through.

One point remains: the presentation should start a conversation: the sooner, the better.

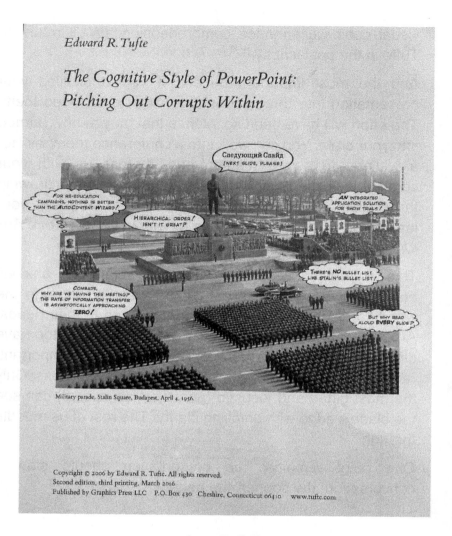

Image #15

One might want to read Edward Tufte's *The Cognitive Style of PowerPoint: Pitching Out Corrupts Within* (Image #15), a searing indictment of mindless slide presentations ($7 in paperback from Amazon):

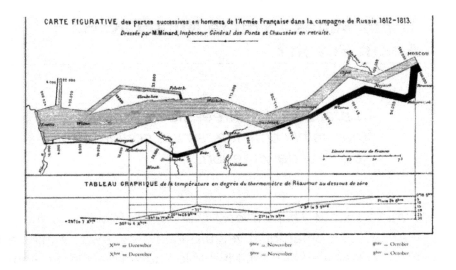

Image #16

Tufte's seminal work, **The Visual Display of Quantitative Information** (Image #16) ($29.62 in hardcover from Amazon), includes this celebrated chart which tracks Napoleon's ill-fated march into and back out of Russia during the abominable winter of 1812–1813:

The chart makes the French Army's unimaginable losses imaginable.

Three Slides? You're Nuts! OK. How About Seven?

In practice, the three-slide pitch may be impossibly concise, so we consider the seven-slide variation.

When I shared *Three Slides Then Shut Up with* my Allegis partners and other local venture investors, I was subjected to a good deal of ribbing. My noble and worthy elders, some former entrepreneurs themselves, questioned my rationality — insisting that it's psychologically and emotionally impossible for entrepreneurs to be boldly concise enough to limit themselves to three slides. How many three-slide presentations had I seen in my years in the Valley? Upon the fourth slide is the presenter sent packing?

I understand very well how difficult it is to restrain yourself when given a chance to present the breadth and depth of the miracle you're but one investment away from realizing. I've been there but robust investment bankers shoved me over the threshold into frugal enlightenment.

Sadly, I have seen less than a dozen three-slide pitches during my VC career, and yet I've been part of the decision to fund many dozens of others. Prolixity isn't an insurmountable obstacle to getting funded. Perhaps I am an irrational and impatient purist, sometimes.

If a terse trio is not workable, seven slides give the presenter more breathing room. The extended template goes like this:

1. *Who we are*. Same as before.

2. *How we'll change the world*. Ditto.

3. *Our market*. This one's new, and here we hit a double bind.

If you position your product as a breakthrough in an established market, investors may worry about the capital required to battle the incumbent giants. But if your pitch proposes a new genre, your audience might question the wisdom of investing in a market with no customers, even though the segment may be poised to explode (of course!).

Your choice in positioning your product, and the response it elicits, will sort the visionary sheep — investors who merely follow — from the ones who believe in technology and have faith in entrepreneurs who create new markets. Huge returns are only made by catching a nascent wave at just the right angle (the product/technology) with the perfect surfboard (the founding team). Uber and Airbnb come to mind.

But no matter how ardently an investor believes in technology, you cannot afford to assume blind faith. As retroactively obvious as Uber and Airbnb are now, those companies demanded a tremendous amount of trust from investors. It's the job of the pitch to engender this trust.

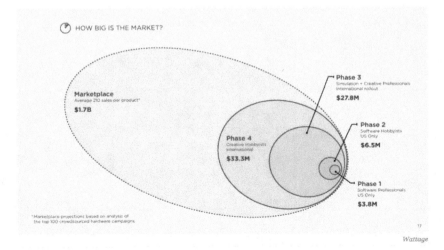

Image #17

You could present your Fortune Teller vision like this beautiful, albeit ill-fated Wattage pitch deck (Image #17).

(I prefer the more traditional lower-left-to-upper-right to signal a bright future, but I'm an irrational purist.)

    4. *Our Competition*. Another new one.

Don't tell us you have no competition. It's a bad way to start a conversation, and there's *always* competition — even if you're simply competing with totally different ways for consumers to spend money.

But please spare us from charts like this (Image #18):

Competitive Analysis				
	Us	Leo	Meg	Mark
Ease of Use	◯			
Memory	◯	●		
Speed	◯		●	●
Reliability	◯			
3rd Party Dev	◯			●
Growth Ramp	◯			
Installed Base		●	●	

Image #18

Or the dreaded Gartner Magic Quadrant[60] (Image #19):

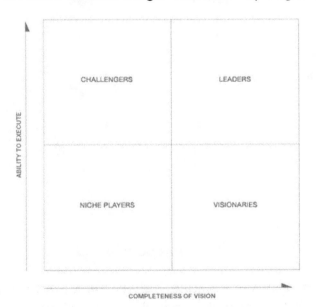

Image #19

[60] https://www.gartner.com/en/research/methodologies/magic-quadrants-research

Just tell us which of your adversaries concern you the most, where they could hurt you, and how you will win.

5. *What you own.*

This can be your IP (patents, trade secrets) or exclusive access to Unobtainium[61] suppliers, but we won't place too much stock in any of that. The most important thing you own is a team of gifted, energetic people: alchemists who will magically transform easily obtained common ingredients. Uber is a good example again: the team built a huge business using technology available to all, including lazy incumbent taxi companies.

6. *The Money Pump.* No change from the three-slide version.

7. The Overview.

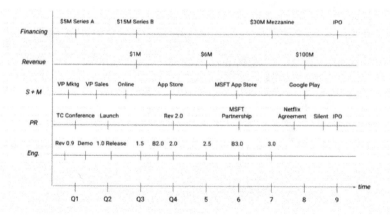

Image #20

This is another Fortune Teller chart. Over three-month increments, it describes parallel company activities ranging from Product Engineering to Financing (Image #20):

---

[61] https://en.wikipedia.org/wiki/Unobtainium

This is more than a lot of noisy detail: it outlines your command of the project's universe. The explicit checkpoints show your willingness to take authority and responsibility. It belongs at the end of the presentation, so it can be left onscreen when you shut up and look at your hands or shoes. The minutiae of schedules, partnerships, and financing plans are yummy bait for questions and arguments.

So, which is it? Three slides or seven?

I still prefer the three-slide version because it's easily memorized and can be delivered on a whiteboard or performed on air guitar. The concluding Overview in the seven-slide format is attractive. It provides a strong natural coda and caesura, but it's much more difficult to recreate on the spot from thin air.

Either way, neither format will induce the First Seventy Minutes of The Hour blight from the previous chapter. Neither provides room to bore your audience by dumping all your knowledge and wisdom on them. You decide.

I'd be remiss if I didn't mention The Demo. If you have a spectacular demo, lead off with it. It will work like the action sequence before the credits in a James Bond movie. Sufficiently adrenaline-rushed, your audience will be ready for calm exposition.

Second, your demo should require little or no explanation. Don't tell us what you're about to do, what you're doing, or what you've done. We'd rather see the demo twice to better get the point than submit to laborious explanations.

A demo that does the selling itself will get you far ahead in the fundraising process. If your demo is a poor salesman, rework it or don't give it at all. An old VC joke contends that there's no better time to raise money than when you have only words — no prototype to break at an embarrassing moment during the demo:

"What does your software run on? PowerPoint."

The Holy Quadruple

A Guide to Clean Thinking — and a BS Detector.

More than three decades ago, I created a variety of Be business plan presentations which failed to follow the rules I'm about to suggest. Aside from embarrassing memories of my own inexperience, my recommendations also stem from participating in hundreds of start-up presentations. Some afternoons, after seeing too many PowerPoint slides, I feel positively Buddhist as I'm dragged back into monkey mind: inside my head I feel a cageful of monkeys shrieking and swinging from the bars. In response I developed an appetite for simplicity and clarity. To soothe myself (and you, I hope), I developed a simple template to analyze what is truly said in these startup pitches. I call it the Holy Quadruple (Image #21):

Image #21

It offers a *canonical* way to present a business to interested parties, to make it understandable and easy to agree or disagree with.

At the top, we find Identity: what the enterprise or product is.

One level down comes Goals: where the business is going, in numbers or other characterizations.

Next come Strategies: paths taken and actions to meet Goals.

Last is the Plan: resources to make everything work (time, money, people, etc.).

This seems banal and straightforward but for two very important elements: the number of words used at each layer and its rate of change. The arrow on the left signals how the number of words and variability increase from layer to layer.

At the top, as the space occupied by the layer suggests, the Identity definition must be concise. If it takes a long time to explain your business, you haven't unearthed its essence. It could also mean there are several businesses here. Which one is the winner we'll want to invest in, and which should be jettisoned or downgraded to support roles for the main act? I once saw an 8-page mission statement at Apple. I called BS — and got my just desserts for not speaking proper Californian. Seriously, though — a verbose ID statement is a bad sign and demands immediate attention.

One layer down at Goals, as the size of that shape suggests, more words are allowed. Even more for Strategies, which require more elaboration. We end with the Plan: the exposé of needed resources — time, people, money, etc.

The rate of change also matters: ID changes very slowly, if ever. Goals can evolve, Strategies change somewhat more frequently with the landscape, and the Plan can and likely will adapt rapidly as market conditions dictate — as competitors make moves.

One may object that my quest for soothing clarity went too far. There are numerous worthy objections to such a simplified approach.

For example, the wall between ID and Goals might not be airtight. "We are what our Goals are," a presenter might object. The same might apply to the boundary between Goals and Strategies: securing a specific retail channel might be.

As a once disorganized presenter and recipient of mounds of business plans, an 'excess' of simplicity can be valuable and welcome. There are always opportunities to add detail later instead of wading into a PowerPoint marsh to find the key ideas of a business.

Lastly, there is the leadership tool besides the Holy Quadruple's BS detection function. Paring the business down to its bones, where it wants to go and through what routes, provide a good way to unite an organization. A strong and simple *"cri de guerre"* binds the group and provides an ever-present compass needle for daily action. A friend insists a business is strong if and only if everyone can recite the company credo, from janitor to CEO, even when drunk and thrown out in the rain by an angry partner, at two AM in one's underwear.

In the real-world seeking simplicity from grown-up companies often disappoints: the result doesn't always speak highly of their leadership or clarity of thought.

Behold a few examples:

> *"Our mission is to empower every person and every organization on the planet to achieve more".*

Or:

> *"Fill the worlds with emotion, through the power of creativity and technology."*

Or, from a company that should know better:

> *"We make technology that enriches people's lives".*

The first example comes from a company which once enjoyed a felicitous accord between name and identity: Microsoft, software for microcomputers — we didn't say personal computers yet. The Redmond giant now presents itself as a gift to mankind, a vague panacea.

One will object that my quest for simplicity is noble but too rigid: Microsoft makes many things other than software for PCs. It makes game consoles, hybrid PC/tablets, cloud software and much more — all producing a market valuation over $2T, often second only to Apple's. Here it's important to distinguish between the company's *essence* — its indispensable core — and what is merely peripheral. Microsoft could be defined more crisply as follows:

> *"Microsoft's core is business software for individuals and organizations."*

Admittedly this would leave the Xbox and other pursuits on the cutting room floor. Validating that definition, though, Microsoft would not die if it stopped making game consoles, but it would if it stopped making Office and Windows software. Microsoft has a mild case of fractured identity, which it has tried to paper over with a bland and all-encompassing motto.

The second example above comes from Sony. Once known as an enviable example of Japanese technology and creativity with Walkman devices, PlayStation game consoles, Trinitron screen technology and plenty more; Sony now encompasses a hodgepodge of activities ranging from digital camera sensors to music, movie studios and financial services. Struggling with another case of fractured identity, the company attempts to cover it all up under a blanket appeal — to emotions, creativity, and technology. Admitting what Sony really is now — a diversified conglomerate — would be less fashionable.

Lastly, Apple, the company which portrays itself as making *"technology that enriches people's lives"*. True as it might be, it doesn't sound as precise and loaded with history as:

> *"We make personal computers, small, medium and large. Plus, objects and services that render those more helpful and pleasant."*

There you have everything in a reasonably concise statement; it covers the main acts, from Watch and iPhone to Mac Pro and the supporting actors tasked with supplementing the lead characters' volumes and margins. Why does Apple, justly renowned for elegant simplicity and sharp marketing, rely on a bland statement which applies equally to so many companies? The answer might lie in its "famously secret" Apple Car project, one which doesn't quite fit my more focused identity. (That being virtuously said, reality tells us the time has come for the Cupertino company to acquire a more fractured identity.)

While the Holy Quadruple surely fails to fully capture all types of business propositions, it offers three important benefits:

As a leadership tool it helps a company develop a shared language to project its identity and future.

The Quadruple is also useful as a BS detector: it sheds a cruel light on companies and people who try to obfuscate, or are too confused for their own good, not to mention ours.

Last it is easy to remember always available for fun at the expense of bloviators, and to the benefit of clear thinkers.

TEAM BUILDING AND UNBUILDING

Transition to HR and Related Topics

While the following chapters are less biographical and more observational on how to make organizations more humane, they are still an important part of understanding my philosophy when contemplating the corporate world and my love-hate relationship with HR.

Firing Well

Ending a work relationship needn't be complicated or traumatic. It can be done safely and respectfully, especially if a clean framework is set up in advance.

Firing is an executive's most important obligation!

Try saying that in polite company, and you're likely to get pained or indignant looks. Behind the strained smiles, you will discern people pondering those well-worn tropes: ruthless executives, short-termism vulture investors or obsequiously scheming human resources minions.

Why? Firing is more important than preaching and maintaining an organizations' Holy Quadruplet — Identity, Goals, Strategies and Plan. In a rational and caring world, questioning the primacy of Good Firing is unfair and dangerous.

Unfair because nothing is more important than the health of a team engaged in the tournament play that most businesses find themselves in. And dangerous because shying away from firing promotes organizational rot and disrespect within the ranks, two hazards which have generated abundant and merciless workplace horror stories.

Some nuggets compare the boss to the comedic spouse who's always the last to know; others sarcastically wonder about the value of working hard when prolonged incompetence and laziness go unpunished; other jibes mock bosses who strut onstage preaching courage in the face of adversity, but endlessly postpone difficult separation decisions.

At HP where I entered the tech world in 1968, I only managed a medium-sized sales team, so I never had to fire anyone — we existed in a culture which only terminated for behavior so blatant that no side ever questioned the decision.

That changed when I was hired to turn around Data General's French subsidiary. Even though the products were less than perfect, the organization's parlous results mostly came from its people. This required painful conversations with bitter individuals, and negotiations made difficult by cultural and legal obstacles. In France, especially in the early seventies, bosses were presumed guilty. Getting someone to leave was painful and required a lot of time and money. This dismayed my American bosses, but our bilingual attorneys made it clear to them that they had let the mess happen in the first place; they added that, contrary to the previous French chauvinistic CEO, I was Americanized enough to be trustworthy.

Dealing with Data General France's sins got me a small 'turnaround guy' reputation. As recounted in chapter "The Exxon Delusion", I was later drafted to put my experience to work as the CEO 2.0 of Exxon Office Systems' French affiliate. When I recounted my Data General adventures to my new boss, Europe General Manager Michel Beurier, he directed me to Gide Loyrette Nouel, a Paris white shoe law firm (now known as Gide). There Partner Bertrand Nouel offered advice, along with a template on a disk, for an airtight negotiated settlement between the company and the person who needed to be let go. Skipping over the convoluted words, the idea was pleasantly simple. The three-step process started with a recognition of harm (real or drummed up for the occasion) inflicted to the employee, such as promises made but never fulfilled. Second the document proceeded to state that the parties had met, negotiated, and agreed upon reparation: several months of salary and a promise to give a good reference. Third it concluded that the individual had had the opportunity to seek legal counsel and pronounced themself fully satisfied by the reparation. As a result, they unequivocally renounced any right to sue their former employer.

There was an additional advantage to the concoction: the monies paid in 'reparation' were, at the time in France, tax-free as compensation for harm inflicted.

It worked, especially because the company let me be generous. We agreed that an avoided lawsuit and an averted demoralizing employee conflict were worth months of salary — six or more — when the company had actually done something stupid.

One day a potentially belligerent employee went to the Labor Inspection Office with the draft agreement I always tendered for calm contemplation and possible legal advice. They were laughed away; told they would likely get a lot less in court and urged to quickly return to the company to sign the agreement before management changed its feebly generous mind.

The process worked, and firmly implanted itself in my mind.

We now jump to Cupertino where in 1985, I was tasked to merge two previously antagonistic organizations (the Apple][and Mac divisions) and lead them to one goal: getting the Mac out of the ditch, following its flamboyant but ineffective début. This entailed taking care of redundancies and other ailments: firing people.

Proud of my Exxon experience in the matter, I tried telling the HR people, in principle staffed to support me, to follow my consensual three-step method. "Oh, no, we have a Process here." The much-abused P-word. "We first have to put people through a Corrective Action Plan and see how they behave." Thankfully a sense of urgency prevailed — the company was in real trouble, so we were allowed to let people go without great ado.

It was sometimes traumatic — in the moment, at least. I still recall an Apple][manager, a former Navy aviator, who regaled us with tales of landing on an aircraft carrier in heavy seas which caused the deck to move up and down by as much as 30 feet. In the final phase, due to the carrier's movement, he would for an instant see his plane pointed at the crew on the underdeck, until the next wave movement lined up the landing strip, mere instants before his plane's hook would grab one of three arresting cables. This was a fearless individual, one deeply committed to a job I was telling him he was about to lose. As we spoke, I saw his emotions cause beads of sweat to instantly form on his forehead, something I had never seen before. He was very civil, and we parted company with best wishes, he for us trying to right the Apple ship and me for his future.

Weeks later I received a postcard from Argentina, thanking me for liberating him from what he had known to be a losing situation. He'd decided to retire, travel to places he wanted to know, and enjoy his new days.

Sadly, not every parting ends up on such a note. Still, that one strengthened my determination to do my best to remove unneeded hard feelings from separation — I mean firing.

Later, when we founded Be, I was determined to handle hiring mishaps without the Apple contortions. I asked the attorney who helped us set up the company if, in California,

I could use the process that had worked well at Exxon. 'Sure, absolutely!' They simply shrugged when I asked about the reasoning behind Apple's ways.

I was set and, as you'll see in the chapter on "Hiring Well", had no trouble with firing during the Be decade.

I feel lucky — when I look around large Valley companies, I rarely see honest and humane terminations. Instead, I see people "getting the message" by being pushed around, sometimes in non-jobs, with HR (Humans Resources) 'counseling' them to look for other opportunities, but not moving decisively. This is demoralizing for the individual and the rest of the organization. Often procrastination comes from higher-ups, senior management, and board members who object to the type of firing I advocate because, they indignantly say, it would reward bad behavior. They prefer the demoralizing spectacle and perhaps to spend money on lawyers, rather than settling with the individual. I don't and I hope you won't.

Hiring Well

Firing Well, discussed in the previous chapter, requires planting seeds of honesty when hiring.

Mentioning firing goes against the rosy mood that often prevails in the mutual courtship ambience of hiring conversations. And yet, when a positive outcome seems likely, I find it a good idea to tell the candidate how I would fire them and why.

For the How, I explain that I go to great lengths to avoid public humiliation. If working together is no longer feasible, I offer a private conversation in which I propose a Covenant Not to Sue. This is a *quid pro quo.* On one hand we offer generous compensation for letting the individual go. On the other hand, the person declares themself satisfied and renounces their right to sue their former employer. For both sides a negotiated solution is much preferable to painful, protracted, and expensive litigation. We'd rather give money (and time) to you and your family than to attorneys. And we let you decide how you announce your resignation. This is preferable to losing sleep, collecting stomach acid, or incurring reputational damage.

Most candidates, when told how they might be asked to leave, shrug: "Of course, glad to hear you have a clean process in place — this is an Employment At Will context, not a union shop." And the conversation can refocus on the job and the candidate's ability to help. But in rare cases the individual becomes agitated and indignant: "That's no way to treat people!" Well, thanks for the warning, we have a culture disconnect, and we know what to do.

Here I won't discuss interview techniques, who should see the candidate, or how to check references. That is all HR territory. I'll just mention a question I like to ask candidates outside the context of the contemplated engineering, sales, or marketing position: "What is the thing, trait or achievement people most misunderstand about you?" It is a dangerous question. It can trigger people into revealing frustrations and resentment they harbor about their past context. Their answer can tell you if the individual comes in with a chip on their shoulder. It is best if the candidate reveals themself to be at peace with the world and, with some humor calmly answers: "No, nothing, I think people generally perceive me fairly." If instead they launch into a recitation of grievances, we get another free warning. (In chapter *New Beginnings: Job offers before starting Be*, where I surveyed my job offers after being fired from Apple, you can see how I applied my own advice.)

In the hiring interview I also dwell on reasons for a possible separation beyond the company falling on hard times. In my book there are two reasons to let someone go: bad attitude and bad judgment. Anything else is repairable. For example, a person with a great attitude but lacking a few technical skills can be moved from engineering to a support function. But a receptionist with bad judgment can do too much damage and needs to go. I know this can sound simplistic or even arbitrary, and relies on executives to have good judgment, which I consider an absolute requirement.

In my experience such a civilized way to set up the right context for separation always works well. In one case at Be we hired an individual who was asked to leave Apple for imprudent online comments. He claimed he had mended his ways, and we hired him on faith — our error. Later, after bouts of increasingly erratic behavior, he brought a boxed tarantula into his cubicle as a pet. He saw my expression and graciously said he got the message, not to worry, he'd see the finance manager immediately and take care of the paperwork.

In another place where I held a board position, I walked into an executive's office and closed the door behind me. The perceptive individual looked at me and asked: "Is this the conversation?" It was. A few weeks later he gave a fine valediction speech at a company meeting, with me sitting in the first-row applauding.

I need to mention one wrinkle in this ideal firing process: The Why question: Why are you firing me? Often people know and don't want to be told. Yet others can only superficially be satisfied by the full choreographed process: they want a last argument. It's best to avoid what can turn into an unnecessarily heated discussion. I take the position that we should focus on the terms and conditions, the choreography, and how the individual can organize an honorable departure. I hold firm: we shouldn't pollute How with a contentious and possibly acrimonious argument about Why. My escape clause is that I promise to hold a frank discussion of Why once every other part of the process has taken place: the Covenant is signed, the checks have cleared, the resignation has been announced, and the individual has held a going-away party. "Come back then, and let's have an appeased conversation." You probably guessed correctly: no one comes back — they're always more concerned with taking a rest, having a good time with the settlement money, and sending out feelers for the next job — nobody really wants to hear an unpleasant or (they feel) unfair story.

As mentioned in the previous chapter, there is still something that needs attention: senior management or board members reacting to the money they may feel is 'wasted'. The question is something like: Why are we giving away so much money, to who you told us was a poor performer, or worse. Couldn't we just terminate them and see if they sue us, and if they do try to win the suit or settlement?

Ah. One must be calm and very diplomatic, explaining that we're merely buying peace, saving attorney fees that would otherwise very quickly exceed any settlement monies, and avoiding possible unrest in the ranks. We're won't look soft — just calm, businesslike, and prepared. In any event can you be sure a lawsuit wouldn't dig up unpleasant truths on our side? The "We shouldn't reward bad performance" irritation, when handled calmly, dies down quickly.

The next chapter concludes my foray into the HR marshes with the oft-dreaded performance review.

The HR-Less Performance Review

After reading this chapter, you will like giving and receiving reviews.

My humane approach to performance reviews stems from more than fifty years of exposure to the process, during which I saw how these appraisals generate anxiety for both sides. The 'reviewee' wonders what unpleasant truths they might hear and how their compensation will evolve — or in more extreme cases whether they will be shown the door. The reviewer often harbors guilty feelings about their management performance, how they helped (or didn't) the individual entrusted to their care. The reviewer might hesitate to give the honest feedback they have failed to deliver in everyday interactions.

It doesn't need to be that way.

After years of struggling with performance reviews, both giving and receiving, I concluded there are only two versions of the anxiety-loaded exercise.

The first version goes head-on into the heart of the matter: "We're satisfied with your performance. As a result, we'll continue to work together. And your compensation is adjusted as follows, in dollars and other instruments."

The individual on the receiving end is anxious and might not hear you well. It is a good idea to calmly repeat your opening statement with a pleasant smile.

A pause lets the news sink in.

Then add this: "Now that we've dealt with the big headlines, I will offer a few comments about your performance, about things you do well and others you could improve upon. But in the end, remember what I said about being more than pleased with your performance and our decision to continue to work together. This means you can feel free to ignore suggestions to improve your game, which we have pronounced 'more than good enough'.

This is unconventional. Yet I find it both more effective and humane. For example, I've always resented performance reviews where the compensation update comes at the end. How do we expect the reviewee to listen to our comments on their behavior on the job with an open mind, while the burning compensation question is still open? Or if our job performance comments are ambiguous and leave the door open to separation? Stating the key review components at the beginning releases tension and allows the individual to listen with what becomes an open mind.

Good, but isn't "feel free to ignore my suggestions" an encouragement to disrespect or insubordination? No, it is simply honest and logical. If, as stated at the outset, we're more than happy with your performance and wish to keep working together, you're welcome to keep behaving as you've done so far and disregard our observations and suggestions.

One will object that this sounds like dry game theory logic. Human interactions don't quite work that way. Most people will not just shrug off suggestions to alter their behavior. Especially if such comments avoid labels such as 'lazy', and instead focus on observations — such as "when you're late to most of our team meetings, it makes us feel disrespected". (Here I veer into the HR vocabulary I wanted to avoid).

It doesn't always work that simply.

Occasionally reviewees argue about facts and consequences. There it's best to stick with the broken record theory, put the tone arm back at the groove's beginning, and restate one's pleased acceptance of performance as it is. We shouldn't get into an old couple's argument: "I want you to stop reading sports news on your phone while we're having breakfast!" "Oh, yes, and what happens if I don't? Will you divorce me?" Herein lies a possible flaw in my optimistic prescription for a painless performance review: what if we're dealing with behavior right on the edge of acceptability? What if you dearly wish for the individual to stop disrupting team meetings with perennially late arrivals?

We're now facing a decision.

One possibility is expressing acceptance of the behavior *for the time being,* explaining that you might not tolerate it forever and could change your position at the next review. I'm not fond of such types of warnings; they cast a pall on the relationship, but it might not be avoidable in the real world.

The other way is the second version of the performance review promised at the beginning. But I dissembled: in my belief system, there is only one humane performance review. If it can't be made to work, we resort to the honest Firing Well method discussed two chapters ago. This is better for all involved, doesn't moralize or criticize, and helps everyone resolve what had metastasized into a hopeless situation. Honesty, a well-written no-fault separation agreement and a sensible amount of money can together restore sanity and dignity.

This is the end of my foray into HR matters. I hope I can save people some of the pain many of us have experienced.

NO FUTURE:
PUNKING LIFE EXPECTATIONS

Apple Now What: Short Introduction

Two broad questions.

In over five decades, every place I worked left a mark on my memory, my behavior, and my worldview. From a health insurance company when at twenty years old to a venture fund in my late sixties, from a chic bar on Deauville's boardwalk to my motherly HP France, all produced feelings of gratitude — for opportunities, learnings, and kind people. But one stands above the rest for its lasting imprint on my life and our family: Apple. I 'only' worked one decade at creation of the "two Steves" (Wozniak and Jobs). But were it not for them, and later for John Sculley's generosity in bringing me to my adopted country, I wouldn't have had the times of my life starting Apple France and then leading Apple engineers in Cupertino when we together got the Mac back on track and improving again, after a spectacular but ultimately disappointing launch. Over forty years after shaking hands with another of my Apple benefactors — Europe VP Tom Lawrence — I still feel a strong interest in the company, mostly like what it does, and of course wonder about its future.

In these final chapters we'll ponder two intertwined topics: the future of Apple's leadership and culture, and the company's Next Big Thing — its next wave of growth now that the smartphone market is reaching saturation.

Apple: Now What? Leadership & Culture

Who will succeed Tim Cook? What will happen to Apple's culture?

Today's Apple bears little resemblance to where I worked from 1980 to September 1990. In fiscal year 1991, Apple's total revenue amounted to $6.3B; thirty years later, in 2021, total sales were over 50 times larger: $365.8B.

Is it the same company?

I can still catch a peek of my first modest office building, still Apple building Bandley 3, on the Cupertino Street of the same name. And not too far away is the circular Apple Park, a Steve Jobs creation visible from space, larger than the US Pentagon. (Ironically and not accidentally, Apple Park stands on land once occupied by HP's now-defunct Computer Division.)

Apple's product lines grew from Macs to smartphones, watches, tablets, earphones, and headphones, as well as a broad array of services: app stores, fitness sessions, financial instruments, music, books and more. But corporate traits last decades. Apple's culture yearns for user-friendliness, beauty, and simplicity. Apple also has an enduring desire to control its destiny — to own the key technologies that enable its products. In tech parlance this is known as "owning the whole stack". Lately that drive has expanded to the strikingly successful development of a line of world-beating microprocessors for Apple phones, tablets and, more recently, Macs, now freed from Intel's lagging, complicated, and power-hungry processor line.

Apple's expanding product lines gives some critics a superficial view of the company's business model, calling it overly complex and dispersed, not to mention overly driven to please Wall Street. Those observers cite the growing role of services in Apple's product array and revenue streams. I hold a simpler view, which is easier to organize and less subject to caprice. I see Apple as a company making personal computers, very small, small, medium, and large: watches, smartphones, tablets and laptop/desktop machines. To me everything else plays a simple supporting role: to either increase or supplement the revenue and profit of Apple's personal computers by making them more helpful, powerful, and pleasant. For example, the millions of apps in the iPhone App Store might be the company's leading services contributor, whose undisclosed revenue was generally expected to exceed $60B in 2021. But the App Store's fundamental *raison d'être* is to give the company's smartphones more power and reach. It works — in fiscal year 2021 the iPhone generated $192B in revenue, more than half the company's total $366B.

Another example combines "owning the stack" and playing a supportive role: Apple Silicon, the range of CPU chips which power iPhones, iPads, Watches, AirPads and more. As one result, music to my old Mac ears, Apple's laptops and desktops are almost completely free of Intel. The success of this strategy is clear in sales numbers: in the December quarter of fiscal year 2022, propelled by Apple Silicon chips, Mac revenue grew 25% versus the same 2021 quarter, while Apple overall only grew by 11%. That trend is likely to continue with more Apple Silicon processors.

Great, but now what? Where is Apple going? Will the tree grow to the sky? Or will the company's size and complexity inevitably doom it to bourgeois mediocrity, stagnating growth and ultimately disruption by new technologies, new market forces management will miss, blinded by its mountains of dollars, or paralyzed by its own complexity?

Examples abound. Once the emperor of enterprise computing, IBM mishandled its entry into the huge personal computer market and is now little more than an undistinguished services company. My alma mater HP failed to survive its founders' departures, and made bad acquisitions and CEO decisions; today, it survives as two companies, HP Enterprises, hard-to-differentiate enterprise computing services, and HP Inc., a supplier of commodity PCs and printers.

For Apple one question agitates the kommentariat: Will it continue to thrive after "losing its soul" when Steve Jobs passed away in 2011? Numbers alone cannot answer that question. Under Jobs' chosen successor, Tim Cook, Apple grew from $108B in revenue in 2011 to $378B ten years later. But skeptics keep objecting that Cook didn't innovate; he just rode the wave Jobs created; he's been a careful warden rather than an innovator. I sometimes feel bad for Tim, whom I like and admire, but who keeps hearing: "Nice numbers, Mr. Cook, but you're no Steve Jobs!" This must be painful, even for someone who learned to steel himself growing up 'different' (gay) in Alabama.

Now in his early sixties, having bought a secluded property near Palm Springs, Cook must think about life after Apple, where he has spent more than a quarter century. A well-run company, Apple has a succession plan, and directors have a slate of potential successors. Jeff Williams is a possibility: well-regarded, he is "Tim Cook's Tim Cook", holding the same Chief Operating Officer job Cook held under Steve Jobs. A possible problem is that he'll soon be a sexagenarian. Other names include Craig Federighi, Senior VP of Software; Eddy Cue, Senior VP of Services; Johny Srouji, Senior VP of Hardware Technologies; and other members of Apple's senior leadership team. On this matter I have but one thought: Apple's next CEO will come from within because its culture would reject a foreign body.

Assuming a smooth "from within" leadership succession, problems remain. Public and not-so-public signs show Apple struggling with the common illnesses which afflict large corporations: internal politics, mendacity, make-work, and myopic attention to process over goals. Call this corporate entropy: As they grow larger and older, organizations inevitably lose the edge that made them successful. Apple's functional organization, an unsung Steve Jobs legacy, avoids the inefficiencies and political strife that afflicted companies such as HP or Microsoft with divisive divisional organizations. On this, see *Vanity Fair*'s "Microsoft's Lost Decade"[62], an account of the evils of inhumane people management practices thoughtlessly imported from General Electric. An ironic paradox is the sluggishness and petty attention to processes injected by an organization intended to support the company's forward movement: HR, Human Resources, recently renamed People. The executive in charge is now called the VP of People, the Chief People Officer, as if a grand name change could solve an existential problem. Internally many complain about protracted hiring processes and the practical impossibility of firing people who need to be let go — who don't fit the organization's needs, goals, or culture. This dangerous tendency is called bozo cancer: unfit people metastasize as they hire more like themselves. I was thinking of Apple's cultural future on a flight to Paris while reading Kelly Johnson's memoir, *Kelly More Than My Share of It All*[63]. He recounts years as head of Lockheed's immensely successful Skunk

---

[62] https://www.vanityfair.com/news/business/2012/08/microsoft-lost-mojo-steve-ballmer

[63] https://www.goodreads.com/book/show/1354758.Kelly

Works, the originator of many iconic planes such as the U-2 and the SR-71. His successor, Ben Rich, also wrote a book titled *Skunk Works*[64], in which he recounts how that successful organization, after success with the F-117 stealth fighter, ended up strangled by paperwork and politics.

In reaction to the threat of rising mediocrity, well-meaning observers have a solution: to stay agile and culturally pure, Apple should split into two or more independent corporations. Such a split could result from antitrust investigators accusing Apple of unlawful excessive power to stifle competition in one or more market segments. The effect on Apple would be profound. Returning to my articulation of the company's business model a few paragraphs above, valuable synergies would be lost. For example, an independent App Store would be forced to support other platforms and operating systems: Android, Windows or whatever Meta unleashes in its quest for Metaverse[65] applications. In this scenario Apple would suffer from no longer enjoying the undivided, synergistic energy of its app stores. This could be the very point of a putative antitrust action: breaking Apple's iron grip and making it easier for others to compete. I feel ill-equipped to speculate further.

But after more than 46 years, the threat of mediocrity doesn't yet appear to have slowed Apple down.

The next chapter will explore the question of Apple's Next Big Thing.

---

[64] https://www.amazon.com/Skunk-Works-Personal-Memoir-Lockheed/dp/0316743003

[65] https://en.wikipedia.org/wiki/Metaverse

What Comes After the iPhone?

The smartphone market is now saturated. Is Apple doomed to incremental growth, or will it find new markets to conquer?

There are three main possibilities for Apple's Next Big Thing: an electric car, AR/VR (Augmented Reality/Virtual Reality) goggles or glasses and healthcare products and services. The first two excite imaginations and constantly feed the Apple rumor mill. The healthcare field is less intoxicating but could offer substantial growth opportunities.

Let's first dispose of the putative Apple Car. It'd be fun to drive a 'better Tesla', elegantly engineered and built, autonomously navigating streets and highways under the control of Apple AI software. But as we now know, true AV (Autonomous Vehicle) technology, also known as Level 5 driver assistance[66], is relegated to an indeterminate future. Tesla has admitted to regulators (while brazenly peddling a $12K "Full Self-Driving" option) that it only offers Level 2 Driver Assistance. Elon Musk's pronouncements aside, there is no known way to safely drive from point A to point B in any environmental conditions without human intervention. I regularly drive my spouse's Tesla S, enjoy the electric experience, and am aware of its limitations.

Setting aside the AV question, I don't see why Apple should enter what is about to become a cutthroat market where dominant Tesla is already battling the likes of Ford, GM, Volkswagen, Audi, Mercedes, BMW, and multiple Chinese

---

[66] https://www.caranddriver.com/features/a15079828/autonomous-self-driving-car-levels-car-levels/

and Korean manufacturers. Existing automakers have infrastructure to build, sell and service cars. Aside from its experience contract manufacturing much smaller objects such as smartphones and other personal computers, Apple lacks the experience or culture to enter the auto industry. Pundits point out that before the iPhone Apple similarly lacked experience in the smartphone industry. Still, the spectacularly successful iPod — which outsold the Mac in 2006 — honed engineering and supply chain management skills which naturally fueled iPhone development and growth. I see no such ramp for an Apple Car.

Amidst regular bouts of Apple Car speculation, a key question is rarely or ever asked: How would an Apple Car make money for the company? Today Apple's gross margin hovers around 54% of Sales — about 40% for hardware vs. over 60% for services. By comparison Ford and GM respectively achieve 15% to 13% gross margins. Focusing more narrowly on electric cars, today nobody aside from Tesla makes money building battery-operated cars. Why would Apple enter such a market? I accept that Apple is working on an electric car project, and respect the intellect and motives of Apple executives, but among all the rumors, where is the Apple Car money story?

The situation is different for Apple entering the AR device field. We got wink-wink references to Augmented Reality at Apple's June 2022 Worldwide Developer Conference (WWDC)[67]persistent and regularly updated rumors, and lately Tim Cook's 'stay tuned' non-answer when queried about the company's AR projects. Culturally AR goggles or glasses would make sense for Apple. The products would leverage Apple's miniaturization experience, advanced

---

[67] https://developer.apple.com/wwdc22/

silicon, well-documented supply chain prowess, sales and service network and software track record. Some even predict a first set of product disclosures within a year or so. This is credible and might occur about when my own words are published.

Here I'll take some risks and ask 'use cases' questions. What would Apple goggles or glasses do for us to trigger another wave of application software development — and revenue? For goggles, playing virtual reality games, and exploration (visiting Mars before Elon Musk gets us there) come to mind. But how often and where would we use them? And at what price, compared against an iPhone? We pay relatively high prices for phones because we use them all the time and everywhere, and because they offer millions of apps. Will that be the case for Apple AR goggles? Recall Steve Jobs' powerful and concise January 2007 positioning of the first iPhone: an Internet navigator, a music player, and a phone. Such a memorable and desirable use case has not yet emerged for goggles — singular or multiple — even after Magic Leap, Microsoft's HoloLens, and Facebook's early explorations.

In many respects, the same holds true for glasses, with additional challenges such as vision correction as they'll be an alternative to our ordinary spectacles, social awkwardness if such glasses include one or more cameras, and more miniaturization and power consumption obstacles. Because they are worn more or less continuously, by filtering and enhancing the "real world", glasses could offer a wide variety of uses, some currently handled by smartphones. For example, reading and answering messages and mail as we walk down the street, video calls, and discreetly entertaining ourselves in less-than-fascinating business meetings. Navigating streets, museums, remote instructions, and assembly instructions are other opportunities for profitable application development. AR glasses seem more promising than goggles. But for these devices to generate another growth wave and become The Next iPhone, we must answer a question: Will we see as many people wearing smart glasses as we have smartphone users today? Engineering and manufacturing glasses is inherently more difficult than making today's iPhones: more sensors, cameras, vision correction, superimposed imaging, audio processing, and advanced silicon to extract powerful advanced computation from smaller batteries. And operating system software must smoothly coordinate everything, along with development tools for a new generation of apps. In conversation a distinguished Valley engineer sees a tall challenge for Apple Glasses: latency. In movies we have a feeling of smooth continuity because successive images in movement happen 24 times per second. So, with Apple Glasses, if we turn our head, the 'show' must update 24 times a second; fresh, believable augmented reality images must arrive every 40 milliseconds, a challenge for computing and power consumption.

We might see an expensive first release as a platform for bleeding-edge users and application developers. I find myself perplexed. Fifteen years ago, no one predicted the iPhone generating hundreds of billions in revenue. Perhaps I am blindfolded by what I think I know, but I still cannot see Apple Glasses reaching the ubiquity of smartphones.

Regardless, Apple Glasses could be a monumental technical achievement. And Apple is a patient company, unwilling to prematurely release unfinished products for publicity.

But there is a third possibility, already visible in the Apple Watch: healthcare. Today we can monitor blood oxygen, sleep patterns, and daily physical activity; we can watch for atrial fibrillation and dangerous falls. Other applications, such as blood pressure or glucose monitoring, dangle larger opportunities — along with technical and legal challenges. For example, I don't see Apple selling a device that involves piercing one's skin, as is the case for most personal glucose monitoring solutions. The opportunity is tantalizing: the US cost of Type 2 diabetes (the more easily preventable kind) was estimated[68] to be more than $360B in 2020 — and rising steadily. Apple is fully aware of that number.

Blood pressure is a valuable health signal that can easily be monitored at home with widely available devices from companies such as Omron[69]. That same company now markets monitoring in a smartwatch which monitors blood pressure from your wrist. But the current device is substantially larger than an Apple Watch and uses an inflatable cuff. Would Apple solve the size and cuff problem, or market a special Apple Watch variant? And deal with FDA registration for a medical device?

Still another application comes to mind: hearing aids. As usual, statistics vary and could be biased. A common one claims that "approximately a third of people in the United States between the ages of 65 and 74 have hearing loss."[70] In other words, "hearing loss would be twice as common as diabetes or cancer." Diminished hearing reportedly leads to cognitive decline and social withdrawal, yet more than a third of people who could benefit from hearing aids don't use them

---

[68] https://www.statista.com/statistics/242157/us-diabetes-type-2--medical-costs-from-2007-to-2020/

[69] https://omronhealthcare.com/products/heartguide-wearable-blood-pressure-monitor-bp8000m/

[70] https://www.healthyhearing.com/report/52814-Hearing-loss-statistics-at-a-glance

because of social stigma and cost. A Jeff Bezos maxim comes to mind: Your margin is my opportunity. For example, Resound[71], a Danish company, markets hearing aids that cost as much as $7K per pair. A smartphone app tunes the hearing experience for circumstances ranging from all-around listening to noisy restaurant scenes to music. Here it's hard not to think of what Apple could accomplish by bringing its successful AirPods expertise to selective intelligent hearing filtering, amplification and directional microphony. The sound quality of streamed audiobooks using Resound devices is unpleasantly inferior to AirPods. There is room for plenty of technology investment between a $249 pair of AirPods Pro and a $7K set of devices — even enough to justify the FDA approval[72] process. Hearing aids don't excite imaginations as much as Apple Cars, smart goggles, or Augmented Reality glasses. But less daunting technical challenges and proven use cases might make the category attractive for Apple to invest in.

Returning to The Next Big Thing, let's consider how the 2007 iPhone came about. Smartphones already existed; BlackBerries, Treos, Nokia devices and more were propelled by revolutionary cellphone networks. Apple was enormously successful with the iPod portable music player, and with iTunes[73] music distribution "by the slice". In 2006, a year before the iPhone, the iPod outsold the Mac. 2006 was effectively a dress rehearsal for the iPhone launch, as 2001 iTunes distribution of digital music prepared the way for 2008 App Store distribution of different digital files: iPhone apps. So, the device that led some to call Apple "The iPhone Company" didn't emerge spontaneously from a vacuum.

---

[71] https://www.resound.com/en-us/hearing-aids/resound-hearing-aids

[72] https://www.fda.gov/consumers/consumer-updates/it-really-fda-approved

[73] https://en.wikipedia.org/wiki/ITunes

And waves such as the PC, the Web and smartphones don't happen by dictum; they arrive when something in the air crystallizes into a category that 'suddenly' aggregates technologies, ideas, habits, and money. For example the PC revolution started years before the Apple][, but it took IBM's 1981 entry into the genre to make it official in offices.

Observers sometimes argue that Apple never invented anything[74]. In their view the company just jumped into and improved existing product categories. Aside from PCs and smartphones, they call the iPad a mere derivation of previous tablet inventions, ranging from Alan Kay's Dynabook concept to Microsoft's Courier, to say nothing of Apple's original tablet-size Newton concept — which preceded the actual shipping MessagePad[75]. None of these efforts bore fruit. Either some ingredients were wrong — such as handwriting recognition, or too weak — such as available processors and operating systems. It took Apple three full years[76] from the iPhone 2007 introduction before it could combine its home-grown A4 processor, the iPhone's iOS operating system and the App Store into today's hugely successful phenomenon.

For the creation of Apple's Next Big Thing, a product that would generate a new wave of growth of iPhone proportions, we're left with an unanswerable question: What is in today's air that we don't see, that Apple could crystallize into an iPhone-caliber surge?

Lastly, let's hope Tim Cook continues to ignore the now perennial and unfair You're Not Steve Jobs prattle, and successfully steels himself against a conscious or

---

[74] https://www.theguardian.com/technology/2012/sep/03/apple-invented-anything

[75] https://en.wikipedia.org/wiki/MessagePad

[76] In 2010 Nokia and RIM still vastly outsold the iPhone

unconscious desire to put his name on a product to 'seal' his legacy.

And Now...

Four waves made technology more intelligible and more affordable. The explosive reaction to OpenAI's ChatGPT signals the rise of a fifth transition: AI for the Rest of US.

Over seventy years I have been fortunate, and often deliriously happy, to be carried by four technology transitions. As a young child in a Paris suburb, I built primitive AM radios. The first versions used a primitive semiconductor, a lead sulfide crystal (galena[77]) with a cat's whisker wire — no electricity needed. Soon an engineer great-uncle gifted me Eugène Aisberg's[78] *La Radio Mais C'est Très Simple!*[79] (*Radio? It's Very Simple!*). This moved me to the vacuum tube era. I recall how, helped by a neighbor my age — and his ham radio buff father — I tried to build a set using a 1LN5 pentode[80] that was sold as not requiring the high-voltage supply tubes usually required to heat their filament. Out of pride I declined his competent father's help to wire and check my project and couldn't make it work.

Not much later, in 1955, I was dispatched to the Kreisker, a Roman Catholic boarding school. That solid institution saved me from my convulsed home and provided a solid humanities education. I was introduced to the solid-state era, with older tubes displaced by smaller, more reliable and less energy-hungry transistors. I still recall one evening in the Prefect of

---

[77] https://en.wikipedia.org/wiki/Galena

[78] https://fr.wikipedia.org/wiki/Eug%C3%A8ne_Aisberg

[79] https://www.amazon.com/-/es/Eug%C3%A8ne-Aisberg/dp/210004107X

[80] https://en.wikipedia.org/wiki/Pentode

Discipline's study, holding in my hands one of the first generally available transistors, the OC 71[81]. We knew we were looking at the future. With the benevolent cleric's encouragement, I wired transistor radios, some official and more clandestine. At night in the dorm, I listened to BBC 4 — and marveled at the *"veddy briddish[82]"* pronunciation of *cor anglais* when announcing a Bach opus.

Born there and then, my enthusiasm for what we now call 'technology' would illuminate the rest of my life — and defeat my imagination.

Once out of Bell Labs, transistors moved to California — particularly to what would become Silicon Valley; they coalesced into integrated circuits and gave rise to companies such as Fairchild, National Semiconductors, and Intel. At Intel, Italian scientist, and original thinker Federico Faggin[83] designed a large-scale integrated circuit (LSI), the 4004[84], the first commercially available microprocessor. It arrived in 1971 and used about 2,300 transistors, descendants of the one I saw in 1955.

Meanwhile, after various adventures, I joined HP's French affiliate in 1968. HP was a grand technology company and an indulgent godmother for this agitated, hungry young man, among many others. HP produced forerunners of Personal Computers: increasingly ambitious programmable desktop machines, culminating in the powerful and extremely expensive 9830[85], which cost upwards of $20K in 1971, fully equipped with a printer and hard disk storage. I loved these machines — they were great fun, but only for the technologists

---

[81] https://www.radiomuseum.org/tubes/tube_oc71.htm

[82] https://forum.wordreference.com/threads/veddy-briddish.507156/

[83] https://en.wikipedia.org/wiki/Federico_Faggin

[84] https://en.wikipedia.org/wiki/Intel_4004

[85] https://www.hpmuseum.org/hp9830.htm

in HP's habitual clientele. HP's 16-bit machines couldn't use the emerging 4-bit and 8-bit chips.

About the same time, two young Steves — Wozniak and Jobs — with experience at HP and Atari, came up with a simple design[86] they offered HP (five times, per Woz). Turned down repeatedly but undeterred, we know what they did next: Apple Computer was born in 1976. The Apple I was quickly succeeded by the more manufacturable and better-packaged Apple][[87], introduced in 1977 for $1298. The Apple][used the MOS Technology 6502 8-bit microprocessor — which was inexpensive and, unlike Intel devices, could be purchased over the counter by hobbyists. The 6502 used a modest 3,700 transistors. It arrived in a market already inhabited by machines from Commodore and Tandy, sometimes called the Holy Trinity of Personal Computing (Image #22):

Image #22: By Tim Colegrove — Own work, CC BY-SA 4.0[88].

Personal Computing was born. Finally, the "Rest of Us" had machines we could handle with our minds, arms, and credit

---

86

https://appleinsider.com/articles/10/12/07/apple_co_founder_offered_first_com puter_design_to_hp_5_times

[87] https://en.wikipedia.org/wiki/Apple_II

[88] https://commons.wikimedia.org/w/index.php?curid=79216985

cards. We were excited because it transformed business, education, and private lives. But none of us imagined that billions of PCs would eventually ship (about 300 million in 2022). Nor could we imagine CPUs sporting billions of transistors (20 billion in an Apple Silicon M2 chip). The PC market is now more than saturated, and sales are in decline, but the grand ride lasted more than 40 years.

While PCs soared, another wave was gathering strength, this time in university labs and research centers around the world: the Internet. The original network of networks has expanded to incomprehensible scale and complexity, but it wasn't then something consumers could use. Critically in 1989 at a CERN[89] (European Organization for Nuclear Research) lab Tim Berners-Lee[90] had a thought:

> *"I just had to take the hypertext idea and connect it to the TCP and DNS ideas and — ta-da! — the World Wide Web."*
>
> *[...]*
>
> ***"Creating the web was really an act of desperation because the situation without it was very difficult*** *when I was working at CERN later. Most of the technology involved in the web, like the hypertext, like the Internet, multi font text objects, had all been designed already. I just had to put them together. It was a step of generalizing, going to a higher level of abstraction, thinking about all the documentation systems out there as being possibly part of a larger imaginary documentation system."*

---

[89] https://en.wikipedia.org/wiki/CERN

[90] https://en.wikipedia.org/wiki/Tim_Berners-Lee

HTTP, the HyperText Transfer Protocol, was born. Tim Berners-Lee did his seminal work on a NeXT machine, the favorite Unix machine of many researchers at the time. But, like Unix at the time, consumers and non-technical PC users couldn't use the Internet. It took the web browser to make the Internet what we use today. Building on the 1993 Mosaic[91] browser from Illinois University, Marc Andreessen and colleagues invented Netscape Navigator. Internet and the Web became interchangeable in common parlance, usage exploded, and PCs got a youth serum. Netscape went public, sold itself to AOL — and was displaced by other browsers from Microsoft, Google, Apple, and others. But the browser lives on.

The fourth wave, smartphones, reached new heights much faster than the PC — faster than anything else in tech history. Like their predecessors, smartphones were born as a combination of emerging technologies — notably cellular networks and microprocessors — offering better computing performance while consuming less electricity. There were several false starts and half-starts. Giant cellphone maker Nokia thought it owned the market but underestimated the importance of software, falling to confusion between four incompatible operating systems. After seeing the importance of mobile devices, Microsoft got hopelessly entangled with the failing Nokia smartphone platform and missed an opportunity to torpedo Android at its birth. Imagine how different the smartphone world would be if then-CEO Ballmer had declared the Windows Phone OS free and open like Google's Android, focusing instead on application revenue. This would have worked much better than glibly dismissing Apple's 2007 iPhone.

---

[91] https://en.wikipedia.org/wiki/Mosaic_(web_browser)

Speaking of which, Steve Jobs' next-to-last creation didn't emerge from a vacuum either. 2006 iPod revenue surpassed Mac sales. The iPod's miniaturization achievements and Tim Cook's supply chain management prowess enabled the iPhone's — and Scott Forstall's team managed the (initially disbelieved) feat of shoehorning a version of Mac OS into pockets. The App Store came out less than a year later, and we got *"There's an app for that"* — the slogan which so succinctly characterizes the universality and ubiquity of the smartphone era. Apple benefited enormously from the iPhone, becoming the world's most valuable tech company. Google's Android smartphone OS, wisely marketed as a "free and open" platform (despite real limitations to its freedom and openness), achieved 72% worldwide market share in 2022. Android contributed massively to the expansion — now saturation — of the smartphone market. In 2022 about 1.43B smartphones were sold around the world.

Now many people wonder whether a new technology wave will reanimate the tech world away from its current incrementalism. Meta, *née* Facebook, CEO Mark Zuckerberg has bet on the Metaverse, a new, all-encompassing environment that will virtualize and subsume all activity — from entertainment to shopping to communicating to working to learning. In physics it's called a TOE, a Theory of Everything[92]. Skeptics abound, pointing to billions of early losses for Meta ($13.7B in 2022[93]), but the company is rich enough to keep pushing through, counting on a prize measured in thrilling trillions. Doubters abound, arguing humans might not want to spend much of their lives

---

[92] https://en.wikipedia.org/wiki/Theory_of_everything

[93] https://www.cnbc.com/2023/02/01/meta-lost-13point7-billion-on-reality-labs-in-2022-after-metaverse-pivot.html

in virtualized dehumanized environments. Zuckerberg seems to join Apple's and Tim Cook's philosophy: Invest in the downturn, while fearful competitors starve their own futures.

Speaking of Apple, the tech rumor mill regularly predicts various VR/AR devices will emerge someday from Cupertino. As much as I like and often admire my old employer, I see two problems with Augmented Reality devices creating a new iPhone-caliber growth wave for Apple. First, the putative devices fantasized by the rumor mill are much more complicated than smartphones. Cameras, motion sensors, real-time internal imaging, fast scene recomputing as the user's head moves to avoid motion sickness, more complicated system software, higher computing power, uncomfortable heat — all are required at levels not currently available. We shouldn't underestimate Apple engineers, nor top management's prudent avoidance of pre-announcements. Still, one question remains: are we desperate to use these putative AR devices the way we use today's smartphones: everywhere, all the time, for (almost) everything. I don't believe so.

But there is another possible fifth wave: Generative AI (Artificial Intelligence). 'Generative' means AI which generates content: text, images, and music — so far. Non-generative AI can recognize faces, analyze situations, and react — as in automated driving. At the end of 2022 we saw an explosion of interest in OpenAI's ChatGPT[94] chatbot. It immediately attracted millions of users, probably the fastest-growing app in history[95]. ChatGPT answered questions, wrote verse, explained quantum physics, and wrote code in

---

[94] https://en.wikipedia.org/wiki/ChatGPT

[95] https://fortune.com/2023/02/02/chatgpt-fastest-growing-app-in-history-could-revolutionize-trading/

many languages. To many, yours truly included, this was AI escaping from research labs: "AI for the rest of us". ChatGPT made mistakes, and critics pointed out its broad use of existing research and various dangers but in parallel presented great examples for many walks of life ranging from education to information integrity in media and business and to private life. For example, generative AI could produce interesting art but also create deep fakes, shifting pictures or words in a compromising context. OpenAI is moving rapidly to build resources for its future. First, it's raising billions from Microsoft and other investors, putting some of that money into a startup to use its technology, and rapidly expanding the store of data on which it runs. The GPT in ChatGPT stands for Generative Pre-trained Transformer.

Some call this a Netscape Moment, a series of events that take technology out of research labs and put it into the hands of millions, if not billions, as the erupting product/technology improves and finds more uses.

Some established AI researchers react with a mixture of condescension: nothing original, just good packaging — to promises of doing much better *soon*. Others, such as Google, once considered an AI pioneer, say little but are said to actively look for ways to respond to OpenAI and to catch-up on real-world GAI applications. Meanwhile OpenAI and investor Microsoft tout the improvements OpenAI technology is bringing to products such as Bing search and its Office products. Of course, this constant dynamic and fast-moving field may (and should) make these comments obsolete by the time these lines come in print.

Will this prove to be another tech and PR soufflé, eventually deflated by technical or even challenges? That is always a possibility. Or more positively, like Netscape OpenAI might find itself overtaken by more muscular, better-financed competitors providing more powerful and usable generative AI.

In a way, like Alexis de Tocqueville meets Arthur C. Clarke[96], these waves were, are and will be the democratization of magic.

---

[96] *"Any sufficiently advanced technology is indistinguishable from magic"*

IN CONCLUSION...

Now, you understand my book's title.

At heart I'm a geek, still in awe of technology since the day I first saw a germanium OC 71 transistor in 1955. Software, which I didn't understand, only added to my joy when I timidly wrote my first programs in 1968. Five decades later silicon and software continue to move forward and generate mixed feelings. Initially empowering and liberating, technology is now also viewed as an instrument of oppression and privacy invasion — a tool to disseminate lies and stoke anger. Resigned ambivalence now replaces the naive unmitigated enthusiasm of my youth. Still, as an inveterate geek, I can't help but rejoice at silicon and AI/ML (Machine Learning) breakthroughs.

There is no ambivalence regarding the first part of the title: my gratitude is unmitigated. My debt is immense to the people who loved me, helped me, and sometimes had to tolerate my pinballing — from schools to jobs, from sentimental education to a successful family which is deeply meaningful to me. Thanks also go to some who didn't like me, who opposed me as I came to benefit from the experience, and the insight I gained later, when more at peace with the world. I tried to see their viewpoints which I missed at the time.

As the Fifth Wave comes closer to shore, this aging Grateful Geek is once again excited.

MANY THANKS

As the title implies, this book is filled with much happy indebtedness. Yet there is more.

First there is our friend Michel Serres[97], philosopher of sciences and member of the Academies Française. Across thirty years of Parisian lunches at the *Train Bleu* and Big Sur hikes, he regaled our family with insights on epistemology — and down-to-earth human natures *aperçus* unfit for this family-friendly publication. In our last conversation two days before his death on June 1st, 2019, Michel once again repeated an old injunction: Write a book! Here it is, Michel, with my thanks for the kind impulse.

Next there is our daughter Camp Director Marie, so called for the organizing skills mentioned earlier. As I was dithering after a book-writing false start, she gathered fragments of my writings over the years and commanded me to get back to my writing station. Thank you, Marie, for the helpful crack of the whip.

Finally, the indispensable Sébastien Taveau, a long-time friend who recently published a book felicitously titled "The Delivery Man", the person who breathes real-word viability into innovators' visions. One day over lunch at Taverna, a Greek restaurant in Palo Alto, when Sébastien asked how my own book project was progressing, I had to sheepishly admit I was neurotically stuck. Writer's block, how unoriginal. Sébastien offered his help, pushed me to write more, and

---

[97] https://en.wikipedia.org/wiki/Michel_Serres

took charge of the editing and production process, a complicated one as I had had yet to discover. Sébastien became The Delivery Man for this book as well. Without him Grateful Geek would have ended as an unfinished manuscript, the source of embarrassed if-only regrets. Thank you, Delivery Man Sébastien!

ABOUT THE AUTHOR

Jean-Louis Gassée (born March 1944 in Paris, France) is a business executive. He is best known as a former executive at Apple Computer, where he worked from 1981 to 1990. He also founded Be Inc., creators of the BeOS computer operating system. After leaving Be he became Chairman of PalmSource, Inc. in November 2004 and served as president and CEO of Computer Access Technology Corporation (CATC), a company which made network protocol analyzers, but left within a year (CATC was purchased in fall 2004 by LeCroy Corporation, a competitor). Gassée resurfaced as a general partner at Allegis Capital, a venture capital fund based in Palo Alto, California, from which he retired in 2022.

In 2009 he started contributing regularly to the *Monday Note* blog, a newsletter covering the intersection of media and technology. Gassée's sharp and incisive analyses of his beloved tech industry have attracted many followers, including well-known technologists.

He is also well-known for his famous one-liners, *aka* Gasséeisms.

Gassée holds a Master of Science from the University of Paris and a Bachelor's Degree in mathematics and physics from Orsay University.

He is married to Brigitte and has 3 children, Marie, Sophie, and Paul.

Source: Wikipedia / The Computer History Museum

CPSIA information can be obtained
at www.ICGtesting.com
Printed in the USA
LVHW050326110723
752038LV00002B/206